GAMMA SOLUTION

James R Warren

BLOXWICH
2022

First Published in the United Kingdom in 2003-2022 by Midland Tutorial Productions

First Edition: 15 August 2022

ISBN 978 1 7396296 8 7

Printed and Bound by IngramSpark

Midland Tutorial Productions Publishers
31 Victoria Avenue
Bloxwich
Walsall
WS3 3HS
United Kingdom

GAMMA SOLUTION

**An Application of Metrics
To the Comparison of
Complete Gamma Function
Solution Algorithms**

First Edition

James R Warren

MIDLAND TUTORIAL PRODUCTIONS
BLOXWICH

To The Glory of
The Loving God

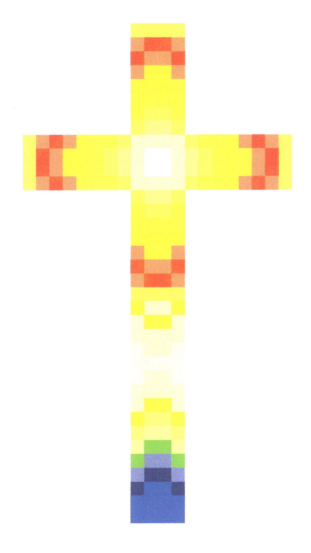

Who Made Our Minds Free

GAMMA SOLUTION
WARREN

PREDICATE

An analytical and experimental study was undertaken in August to October 2004 to assess the degree to which theoretical and experimental code and performance metrics might assist the choice of published Complete Gamma Function (CGF) solution algorithms, or inform the development of existing or new CGF methods.

The results of that research are the subject of this disquisition and are presented and assessed in the foregoing chapters.

The outcomes fulfil this Schedule agreed at The Technology Innovation Centre, Birmingham on 3rd August 2004:-

1   To identify published studies regarding the application of metrics, especially Halstead Metrics, to the computation of convergent mathematical and statistical functions.

2   To identify a selection of algorithms for the computation of complete gamma functions and to program these as functional subroutines.

3   To compare the complexity of the computational functions and calculate Halstead Metrics in regard to Length, Volume, Volumes Ratio and Level.

4   To compute and compare solution times for each compiled gamma function estimator, in the light of algorithm accuracies.

5   To identify and recommend further appropriate metrics, if applicable to gamma functions.

6   To compare, evaluate and recommend algorithm quality metrics applicable to complete gamma function estimation.

Computational experiments were applied using a Mesh Matrix personal computer. Those CGF estimators coded and exercised are listed in Table Fourteen of the Conclusions. Algorismic specifications of these procedures may be read in Chapters Two and Ten. The accuracy of these methods was gauged against the relevant EXCEL® intrinsic function assembly =EXP(GAMMALN(Cell)) that I deemed fiducial.

The speed of these methods was computed using interpreted Visual Basic 6.0 subroutines implemented on the Mesh Matrix platform. Additional timing tariff computations were made in this same medium.

Token counts were made manually using the CGF methods VB6.0 source code.

This document was composited using the MicroSoft OFFICE® suite of canonical applications tools, in particular WORD® (text), EXCEL® (dynamic tabulations) and PhotoEdit® (diagrams).

Some MATHCAD® scratchpad trials were performed.

Most of the journal publications cited are available on the InterNet (at 21 October 2004), and that was used to garner them, where necessary employing Acrobat Reader 6.0 or DVIWIN.

No other platform resources were required or applied.

LIST OF CONTENTS

CHAPTER ONE

Introduction

The purpose of this study is to bring together two things: A mathematical object called a Complete Gamma Function; and an assembly of software metrics. The metrics will enable us to assess the suitability of several different ways of calculating a Gamma Function's value.

What is a Gamma Function?

A gamma function is an important example of a class of generalised factorials used in combinatorics and other statistical applications, for example chi-squared probability computations.

An ordinary factorial is also useful in a wide range of applied mathematical work, but its problem is that it only exists for positive integers; i = 0,1,2, ... , n, where n is some arbitrary positive whole number.

This is because a factorial is defined by the product series:-

$$n! = 1 \times 2 \times 3 \times 4 \ldots \times n = \prod_{i=1}^{n} i$$

Equation 1

One important application of the factorial is in answering the question "How many ways can I choose k objects from n of them, where order does not matter, but repeats are not allowed?".

Technically, this is calculating a Binomial Coefficient, and is very useful in statistics and its dependent sciences and technologies.

The answer is given by[1]:-

$$^{n}C_{k} = \binom{n}{k} = \frac{n!}{k!(n-k)!}$$

Equation 2

For example, if we choose k = 3 of n = 5 objects, call them A, B, C, D and E, then $^{n}C_{k}$ = 10 distinct triples. We can elaborate these choices as: ABC, ABD, ABE, ACD, ACE, ACB, ADE, AEB, AEC and AED.

As aforementioned, n! is only good for integers and we would like a function versatile enough for fractional numbers too.

In fact we need to think about two sorts of fractions at this time: Rational Fractions such as $^{1}/_{7}$ or $3^{1}/_{7} = {}^{22}/_{7}$ which have integers as both dividend and divisor. These will form recurring decimals such as 3.142857142857; and secondly, a special kind of non-rational fraction called a Transcendental Number.

Transcendental Numbers are troublesome but exciting. Their decimals never recur with any perceptible pattern *and never terminate*. This means we can never compute a transcendental number *exactly*, only approximately, though that approximation may be good to a thousand places of decimals.

The most famous example of a transcendental number is The Ludolphine Constant, π, which is about 3.14285714286, when my Casio fx-992x truncates it to the first eleven decimal places.

You can never compute a transcendental number with a nice fixed-length analytic formula like $E = mc^2$. The best you can do is approximate its value with some numerical recipe called an algorithm, in particular with convergent summative or product series. Or you can estimate it with some ad hoc rule.

A Gamma Function is a transcendental number.

You can feed a Complete Gamma Function algorithm with a single input argument that can be an integer, a rational fraction or an irrational, *but the output resulting Gamma is always transcendental* even when it is masquerading as a whole number.

The formal definition of a Complete Gamma Function is:-

$$\Gamma(z) = \int_0^\infty t^{z-1} e^{-t} dt$$

Equation 3

Because $\Gamma(z)$ is transcendental there is no analytic solution to this integral.

You could of course estimate the value of $\Gamma(z)$ for any real positive z by substituting some big number for ∞ (infinity) and applying a suitably-chosen numerical quadrature. But that would be very costly in computer time and the end result would probably be imprecise.

It is much better to use some of the other tricks we will review.

By the way, for integers, ordinary factorials and gamma functions are related by the expression:-

$$z! = \Gamma(z + 1)$$

Equation 4

So $\Gamma(6) = 5! = 1 \times 2 \times 3 \times 4 \times 5 = 120$

and both grow very quickly. 3! is only 6 but 63! is $1.9826083154 \times 10^{87}$, a figure about eight and a half billion times the dollar size of the American gross national product.

But $\Gamma(1/3)$ is also defined, as is $\Gamma(1/2)$ which is $\pi^{1/2}$, showing how the Gamma of a rational fraction is transformed to a transcendental.

Why Know this?

Without the Gamma Function it is impossible to define (or compute) the values of several statistical intermediates.

For example, we may use Chi-squared tests to determine the likelihood of differences between two sets of data being real or merely apparent. Perhaps groups of control and treated patients in drug tests; or before and after profiles of one group of patients.

The Chi-squared probability function is defined as:-

$$P(\chi^2|v) = \left[2^{\frac{v}{2}}.\,\Gamma\left(\frac{v}{2}\right)\right]^{-1} \int_0^{\chi^2} (t)^{\frac{v}{2}-1}.\,e^{-\frac{t}{2}}.\,dt$$

Equation 5

as you can see, without $\Gamma(v/2)$ there is no access to knowledge of $P(\chi^2 \mid v)$.

There are many other statistical functions, including Beta, t-quantile and allied Eulerian Functions from which the Gaussian Distribution and other industrially-useful objects may be derived[2].

In advanced technology, Gamma Functions are applied in the computation of harmonic oscillator characteristics for applications in chemical quantum mechanics[3]. Or in the modelling of airplane glass failure times, as researched by Ed Fuller in 1993[4].

It is very fulfilling to discuss with you fine and useful things like the Gamma Function. Things that could make a real difference to the way people live.

And yet we should pause to remember that Leonard Euler, James Stirling and the other eighteenth-century Enlightenment giants who brought Gamma to birth did not want or look for these splendid applications. Their only motives were the love of art and the celebration of God's Providence.

CHAPTER TWO

Some Methods of Gamma Function Estimation

Ask any layman how many centers there are in a triangle and he will briskly reply "one, of course". But mathematical theoreticians say they have proven that each triangle has as infinite number of distinct centers, and I myself have a working knowledge of five. It all depends upon what you call a "center", and why.

Because there is no determinate way of calculating a Gamma Function, I surmise there are an infinite number of computational methods thereto. At any event, I have seen a couple of score in papers and handbooks. Inevitably, we will consider a tiny subset of these known procedures, compare them with imperfect yardsticks, and know that our best specimen cannot be the paragon.

Stirling's Formula[5,8]

In his 1730 treatise *Methodus Differentialis*, the Scottish mathematician James Stirling (1692-1770) proposed an elegant approximating function for the factorial, handy when treating of large numbers:-

$$n! \approx \sqrt{2\pi n}.\left(\frac{n}{e}\right)^{n}$$

Equation 6

This neat formula is general enough to furnish a basis of Gamma estimation, re-arranged as:-

$$\Gamma(z) = \sqrt{2\pi}.e^{-z}.z^{z-\frac{1}{2}}$$

Equation 7

For smaller arguments it is worthwhile to apply a corrective series expansion as[11]:-

$$\Gamma(z) = \sqrt{2\pi}.e^{-z}.z^{z-\frac{1}{2}}\left[1+\frac{1}{12z}+\frac{1}{288z^{2}}-\frac{139}{51840z^{3}}-\frac{571}{2488320z^{4}}+\ldots\right]$$

Equation 8

By taking logarithms of both sides it is possible to clarify this adjusted estimator as:-

$$\log_{n}\Gamma(z) = \left(z-\frac{1}{2}\right).\log_{n}z - z + \frac{\log_{n}2\pi}{2} + \sum_{k=1}^{m}\frac{B_{2k}}{z^{2k-1}(2k-1)2k}$$

Equation 9

where B_{2k} is a Bernoulli Polynomial calculable via binomial expansions. A list of interesting B_{2k} is provided in Appendix A.

It turns out that after B_6 the absolute values of the Bernoulli Polynomials *fail to decrease*: Unexpectedly they start to increase, breaking the series convergence and making Stirling's Formula a *worse* estimator when expanded beyond that term.

So *used uncritically* Stirling's Formula is a false friend.

A detailed examination of Bernoulli Numbers may be consulted at Smith[17].

But with due prudence we can exploit it as a sufficiently-accurate tool for large z's, or as a starting point for rapidly-convergent, high-accuracy iterations should suitable such secondary tools come to light.

Also we can substitute or supplement a six-term corrective series with some quick and finite mathematical finial that squeezes a few extra decimal places of precision out of the formula, at least in some restricted range of z.

An early example of such an "enhancement" is Burnside's Formula:-

$$x! \approx \left(x + \frac{1}{2} \right)^{x+\frac{1}{2}} . e^{-x-\frac{1}{2}} . \sqrt{2\pi}$$

Equation 10

where x is understood to be a positive real.

More recently, Windschitl[13] extended Stirling's original formula as a power series $Log_n M(x)$ and noted an analogy of the expansion of $Exp(2 Log_n M(x))$ with the expansion of $x.Sinh(1/x)$.

This led him to compose:-

$$\Gamma(z) \approx \sqrt{\frac{2\pi}{z}} \left[\frac{z}{e} + \sqrt{z.Sinh\left(\frac{1}{z}\right) \left[\frac{1}{810z^6}\right]} \right]^x$$

Equation 11

This is said to better eight digits of precision when x>8, but its economy does of course depend upon a fast and accurate Sinh routine: There is of course no finite, analytic expression for the hyperbolic sine, either. (Equation 11 is given incorrectly in Reference 8).

A few years ago I too tampered with the Stirling Formula to obtain[6]:-

$$\Gamma(z) \approx \sqrt{2\pi}.z^{z-\frac{1}{2}} . e^{-z} \left(1 + \frac{1}{8z} \right) \left(1 - \frac{1}{8\pi z} \right)$$

Equation 12

which has a 1.3% relative error at $z = \frac{1}{2}$ and $2 \times 10^{-5}\%$ in the region of $z = 4.15$.

The next year[7], I stabilised this estimator in the range $z = 1$ to 12 by the assembly of Equation 12 with a suitable least-squares cubic fitment to form:-

$$\Gamma(z) \approx\ ^3P(z^{-1}) \cdot \sqrt{2\pi} \cdot z^{z-\frac{1}{2}} \cdot e^{-z}\left(1 + \frac{1}{8z}\right)\left(1 - \frac{1}{8\pi z}\right)$$

Equation 13

Appropriate coefficients are listed in Appendix A.

The Lanczos Approximation[8,12]

One high-precision approach to Gamma computation is the Lanczos Method with a claimed absolute error $|\varepsilon| < 2 \times 10^{-10}$ for any complex z for which $\Re z > 0$.

The relevant equation (technically of course an inequality of approximation) is:-

$$\Gamma(z) = \left[\frac{\sqrt{2\pi}}{z}\left(p_0 + \sum_{n=1}^{6} \frac{p_n}{z+n}\right)\right](z+5.5)^{z+\frac{1}{2}} \cdot e^{-(z+5.5)}$$

Equation 14

The seven p_i are given in Appendix A.

Reference 12 is "Numerical Recipes in C" in which a Lanczos Module is published.

Some Classical Approaches

Strictly speaking, Gamma is seldom if ever defined by convergent series: Gamma Series are asymptotic. Alarmingly, this means that at some remote juncture expansion terms may actually *diverge*, as we have recently seen to apply to Stirling's Series: *Caveat Emptor*.

Many sum or product series expansions for Gamma exist in the literature and accessible examples can be seen in Abramowitz and Stegun[11], CRC Manuals, and on the Net at, for example, Wolfram Research, where derivations are discussed under "Gamma Function"[9] and selected series expansions listed under "Gamma"[10]. The Wolfram Research sites content is obtainable as a codex book called "CRC Concise Encyclopedia of Mathematics"[15], which is optionally supplemented with a CD-ROM.

More specialised or more recent Gamma approximators abound in the journal literature and in specialist books. Some of the recent formulations for high-precision programming may be seen at Winitzki[18].

These methods differ widely in their speeds of convergence, delayed divergent behaviour, vulnerability to errors of rounding and truncation, and the need for ancillary subfunctions. But their stereotyped and extensible characters tend to limit complexity and increase flexibility. All yield up definitive precisions, given enough iterations. And all make a limited appeal to awkward constants.

We will discuss a few key forms.

The Euler Infinite Product

This is defined as:-

$$\Gamma(z) = \left[ze^{\gamma z} \prod_{r=1}^{\infty} \left(1 + \frac{z}{r}\right) e^{-\frac{z}{r}} \right]^{-1}$$

Equation 15

Clearly, we are not going to elaborate an infinite number of multiplications in a calculation, so in practice we will substitute ∞ with some big positive integer, depending upon the desired accuracy of the result, and in the light of the size of z.

γ is the Euler-Mascheroni Constant. This object can itself be computed as a one-off using convergent series and has a value near to 0.5772156649. High-precision γ is treated in the anonymous but well-resourced paper Reference 16.

Exponentiated Summation

The Euler Product can be manipulated to give a summative form that computes reciprocal Gamma:-

$$\frac{1}{\Gamma(z)} = z.\exp\left[\gamma z - \sum_{k=2}^{\infty} \frac{(-1)^k \zeta(k) z^k}{k} \right]$$

Equation 16

This implicates the Riemann Zeta Function, $\zeta(k)$, which is fortunately simple and rapidly convergent for large arguments:-

$$\zeta(s) = \sum_{k=1}^{\infty} k^{-s} = \frac{1}{s-1} + \sum_{n=0}^{\infty} \frac{(-1)}{n!} \gamma_n (s-1)^n$$

Equation 17

for $\Re s > 1$.

Or, in the idiom of Equation 16:-

$$\zeta(k) = \sum_{i=1}^{\infty} i^{-k}$$

Equation 18

for k = 2,3, ... , ∞.

Note that γ_n is given by:-

$$\gamma_n = \lim_{m \to \infty} \left\{ \sum_{k=1}^{m} \frac{(\log_n k)^n}{k} - \frac{(\log_n m)^{n+1}}{n+1} \right\}$$

Equation 19

The Bourget Asymptotic Series

This expresses reciprocal Gamma as a simple algebraic polynomial:-

$$\frac{1}{\Gamma(z)} = \sum_{k=1}^{\infty} a_k z^k$$

Equation 20

where the coefficients satisfy the recurrence relation:-

$$a_n = na_1 a_n - a_2 a_{n-1} + \sum_{k=2}^{n} (-1)^k \zeta(k) a_{n-k}$$

Equation 21

The first twenty-six a_k are listed in Appendix A as Davis Series c_k. The asymptotic series expansion is general to any absolute value of z.

Hastings Formulas[19]

In the early fifties of the last century, C Hastings provided some Gamma approximators that were also straight-forward algebraic polynomials.

However, in these cases there is no claim of analytic extensibility and indeed the two formulae are only good in the range $0 \le x \le 1$.

In order to access the gamma values for larger arguments we would be expected to exploit the recurrence formula:-

$$\Gamma(z+1) = z\Gamma(z)$$

Equation 22

The Hastings Type I formula is:-

$$\Gamma(x+1) = 1 + \sum_{i=1}^{5} a_i x^i$$

Equation 23

which exhibits an absolute error $|\varepsilon(x)| \le 5 \times 10^{-5}$ in the given interval.
Hastings Type II is:-

$$\Gamma(x+1) = 1 + \sum_{i=1}^{8} b_i x^i$$

Equation 24

which decreases the absolute error to $|\varepsilon(x)| \le 3 \times 10^{-7}$ in the range $0 \le x \le 1$.
The requisite coefficients for both estimators may be found in Appendix A.

CHAPTER THREE

Some Empirical Approaches to Method Accuracy

Two desiderata inform our quest for the best Complete Gamma Function algorithm over a given interval: Speed and Accuracy.

Such key needs as least Cost, least Storage and other economic requirements are subsumed, at one remove or another, by process Speed.

On the other hand, such requirements as Reliability, Robustness and Utility come under the general ambit of accuracy.

Speed and Accuracy combine, so to say, in the phenomena of series' convergence.

Some Accuracy Measures

Statistical Measures

Among the elementary statistical measures of relative accuracy three metrics are immediately suggested:-

(a)   The Root Mean Square, RMS

(b)   The Standard Error, ε

(c)   The Population Arithmetic Mean, μ

These statistics are not orthogonal, and in the context of our study are related by the parameter Solutions Difference, δ:-

$$\delta = \Gamma_{test} - \Gamma_{fid}$$

Equation 26

where Γ_{test} is a Gamma Value computed by a Method under Trial and Γ_{fid} a Gamma Value at the same z as yielded by a chosen Fiducial Gamma method. In this Chapter, the Fiducial Method is that furnished by the EXCEL® intrinsic assembly =EXP(GAMMALN(Cell)). In other trials not involving this spreadsheet tool we may of course select a fiducial method that we know, or think, will yield up sufficiently-accurate estimates in the given z range.

The Mean Difference, $\mu\{\delta\}$, is given by:-

$$\mu\{\delta\} = \frac{\sum\left(\Gamma_{test} - \Gamma_{fid}\right)}{n}$$

Equation 27

where n is the *Number of Range Data-Points* (i.e. the number of intervals plus one).

In order to examine the relations between RMS, Standard Error and Mean it is convenient to develop the Sum of the Squared Deviations about the Mean:-

$$\sum \left(\delta - \mu \{\delta\} \right)^2 = \sum \left(\delta - \mu \right)^2$$
$$= \sum \left(\delta^2 - 2\delta\mu + \mu^2 \right)$$
$$= \sum \delta^2 - 2\mu \sum \delta + n\mu^2$$

Equation 28

The Population Variance of the Differences is defined as:-

$$V = \frac{\sum \left(\delta - \mu \right)^2}{n} = \sigma^2$$

Equation 29

where σ is the Population Standard Deviation.
Substitution of Equation 28 in Equation 29 then gives:-

$$V = \frac{\sum \left(\delta - \mu \right)^2}{n}$$
$$= \frac{\sum \delta^2 - 2\mu \sum \delta + n\mu^2}{n}$$
$$= \frac{1}{n} \left(\sum \delta^2 - 2\frac{\sum \delta}{n} \sum \delta + n\mu^2 \right)$$
$$= \frac{\sum \delta^2}{n} - 2\mu^2 + \mu^2$$
$$= \frac{\sum \delta^2}{n} - \mu^2$$

Equation 30

Or equivalently:-

$$\sigma^2 = \frac{\sum \delta^2}{n} - \mu^2$$

Equation 31

Now by definition:-

$$RMS = \sqrt{\frac{\sum \delta^2}{n}}$$

Equation 32

and:-

$$\varepsilon = \frac{\sigma}{\sqrt{n}}$$

Equation 33

Therefore:-

$$\frac{\sum \delta^2}{n} = RMS^2$$

Equation 34

and:-

$$\sigma^2 = n\varepsilon^2$$

Equation 35

Substituting these last two in Equation 31 and re-arranging we get:-

$$RMS^2 = n\varepsilon^2 + \mu^2$$

Equation 36

This allows us to identify RMS as a Pythagorean Modulus (in geometrical terms a "hypotenuse") to the Standard Error and the Mean.

We shall choose to focus on the RMS and compare it with the Minima and Maxima of $d\%_s$, a metric that we shall now discuss.

The Specific Defect

The Specific Defect is a measure of the *relative error* in the computation of an estimate.

In our context, the Specific Defect, d_s, is:-

$$d_s = \frac{\Gamma_{test} - \Gamma_{fid}}{\Gamma_{fid}} = \frac{\delta}{\Gamma_{fid}}$$

Equation 37

It is capable of being positive or negative, but, except in the cases of catastrophic method failure, should be much smaller than unity: And the smaller its absolute value the better.

Specific Defect is convenient for comparisons at individual measurement points; or between like unitary statistics such as means.

In some circumstances, especially manual data inspections, we may prefer a percentage expression of the Specific Defect:-

$$d\%_s = 100d_s$$

Equation 38

The Mean Equivalent Figure, MEF

The Equivalent Figure, E, is the count of leading digits that are identical between two compared numbers.

For example, if we compare 91.6854762 with 91.6864762 the E is 4, because the fifth digit differs between the numbers: And only one digit *needs* to differ to break the matching. The E of 0.916864762 and 0.916864293 is six: Leading and trailing zeros do not count. In this last case the Matched Common Value is 0.916864.

This shows that the concept of E differs subtly from that of Significant Figures of Accuracy, SFA. The E matched value is determined by *truncation*, not *rounding*.

Notwithstanding this, given a sufficient number of compared value couples, we may define a Mean Equivalent Figure, MEF, as the mean digit count in the matched values. Specifically:-

$$MEF = \mu\{E\} \approx -\mu\left[\log_{10}\left|d_s\right|\right]$$

Equation 39

And Equation 39 will of course function on any chosen radix.

MEF is a handy intuitive measure of the average accuracy of a given technique. Except in pathological catastrophic failures of computational algorithms (which MEF is especially good at betraying), MEF is always positive; always limited by the inherent fiducial precision (e.g. fifteen figures for double-precision denary); and otherwise the bigger the better.

The Roster of Methods

Table One lists the EXCEL® worksheets used to compare the accuracy of four families of Complete Gamma Function computational methods in three distinct ranges of the argument z.

The methods, based upon the respective equations of Chapter Two and the Appendix A coefficients, are:-

    A   Classical Group
            (i)   Exponential Summation, ES
            (ii)   Euler Product, EP
            (iii)   Davis Series, DV
    B   Lanczos Group
            (i)   Lanczos Approximation, LA
    C   Fitted Polynomial Group
            (i)   Hastings Type II, HT
    D   Stirling Group
            (i)   Stirling's (1730) Formula, ST

(ii)   Stirling's (Bernoulli) Extended Formula, SE

(iii)   Warren's Method Type I, WO

(iv)   Warren's Method Type II, WT

(v)   Burnside's Formula, BS

(vi)   Windschitl's Formula, WS

The full range of z between $0 \leq z \leq \infty$ yields a "bathtub" curve for $\Gamma(z)$, but $\Gamma(z=0) = \infty$ and $\Gamma(z=\infty) = \infty$.

Accordingly, Range Three between $0.0001 \leq z \leq 1$ describes a plunging monotonic curve in $\Gamma(z)$ until $\Gamma(1) = 0$.

Range Two, $1 \leq z \leq 2$ extends between $\Gamma(1) = 0$ and $\Gamma(2) = 0$ with a minimum at about $\Gamma(1.46) \approx 0.885604$.

The graph of Range Two is qualitatively an asymmetrical slight sag.

Range One, $1 \leq z \leq 12$ overlaps Range Two, which is wholly included in Range One. The curve of Range One is a sharp monotonic climb from $\Gamma(1) = 0$ to $\Gamma(12) = 11! = 39,916,800$.

Code	Description	Range Class	Range Lower	Range Upper
M	Range Definitions			
Z	The Zeta Function			
ES3	Exponentiated Summation	3	0.0001	1
EP3	Euler Infinite Product	3	0.0001	1
DV3	Davis Series Solution	3	0.0001	1
LA3	Lanczos Approximation	3	0.0001	1
HT3	Hastings Type II	3	0.0001	1
ES2	Exponentiated Summation	2	1	2
EP2	Euler Infinite Product	2	1	2
DV2	Davis Series Solution	2	1	2
LA2	Lanczos Approximation	2	1	2
DV1	Davis Series Solution	1	1	12
DV1A	Davis Series Solution	1	1	12
LA1	Lanczos Approximation	1	1	12
ST1	Stirling Formula	1	1	12
SE1	Extended Stirling Formula	1	1	12
WO1	Warren Method Type I	1	1	12
WT1	Warren Method Type II	1	1	12
BS1	Burnside Formula	1	1	12
WS1	Windschitl Formula	1	1	12
GC	General Constants			
SC	Stirling Coefficients			
WC	Warren Coefficients			
LC	Lanczos Coefficients			
DS	The Davis Series			
HC	Hastings Coefficients			

Table 1

No one algorithm yields a good gamma in all ranges, and as a rough rule Classical Group methods are strong at low arguments and Stirling Group methods at high z's.

Many widely-published techniques are also notably mediocre, even for "sympathetic" arguments in Range Two.

The Davis Series method, which gives very precise $\Gamma(z)$ over a wide range in my JavaScript research toy GAMMATOY.HTM (http://www.jamesrwarren.com), fails utterly for z > 3.5 when implemented in EXCEL®. This is despite the notional 15-figure accuracy of both

JavaScript and EXCEL, and formulaic re-arrangements of the EXCEL schemes (DV1 and DV1A) achieve no significant improvement.

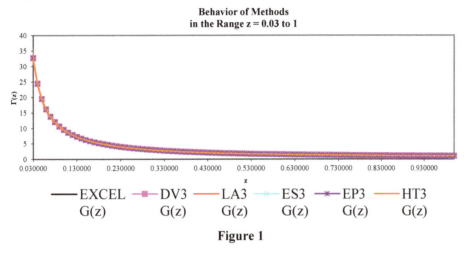

Figure 1

Figure One illustrates most of the Range Three behaviour. To the precision of inspection, all five Range Three methods returned respectable estimates throughout the range.

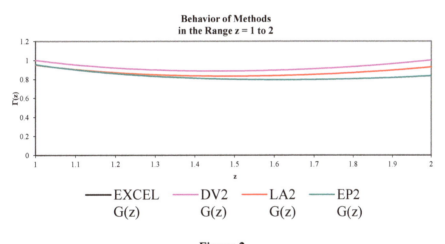

Figure 2

The three Range Two methods tested: Davis, Lanczos and Euler Product diverged. The Davis Series followed the fiducial curve but Lanczos and Euler Product both significantly underestimated gamma.

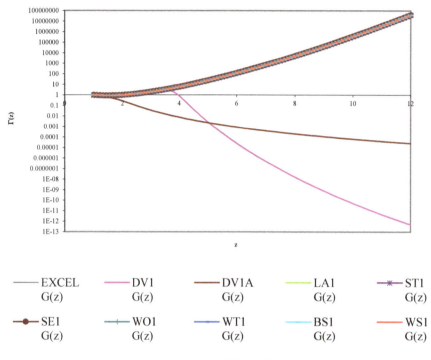

—— EXCEL	—— DV1	—— DV1A	—— LA1	—✳— ST1
G(z)	G(z)	G(z)	G(z)	G(z)
—●— SE1	—+— WO1	—— WT1	—— BS1	—— WS1
G(z)	G(z)	G(z)	G(z)	G(z)

Figure 3

The last of these qualitative plots, Figure Three shows Lanczos and Stirling Group methods faithfully returning good gammas in the presence of early dramatic failures of both Davis Series arrangements.

Relative Errors Analysis

A more revealing statistical analysis may be reviewed in terms of the Solutions Difference RMS and the Percentage Specific Defect extrema for each tested method.

These and other accuracy statistics are tabulated in Appendix B.

Almost all methods exhibit RMS values at one or other of the d‰s extrema.

As a rough rule, methods which work properly in Range Three (i.e. the lowest z's) have RMS at or near the defect minimum, whilst Range One methods have RMS near their defect maxima. Range Two methods can go either way.

Davis Method applications always show wide defect ranges balanced equally around zero, but this is no guarantee of utility as we have seen.

RMS errors tend to be very low for Classical approaches in Range Three and high in Range One. In Range One Stirling Group methods show relatively narrow defect bands reflecting their superior consistency of performance, at least in the higher reaches of z values.

The Lanczos Method is in these terms both accurate and precise for Ranges Two and Three, but a poor performer in Range One ($1 \leq z \leq 12$).

Figure Four shows that overall the Davis Series method wins decisively in Range Three, but has a wide defect latitude whilst Lanczos is mediocre but consistent. The Hastings Polynomial is as good overall as Exponentiated Summation and exhibits smaller defects.

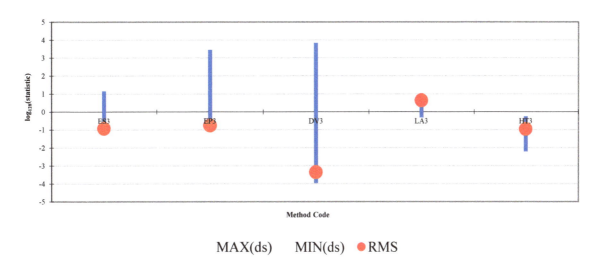

Figure 4

In the central Range Two ($1 \leq z \leq 2$) Exponentiated Summation breaks down completely (probably because of poor zeta definition) but the Euler Product and Lanczos Approximation hold their own with Davis maintaining its lead as the method of choice.

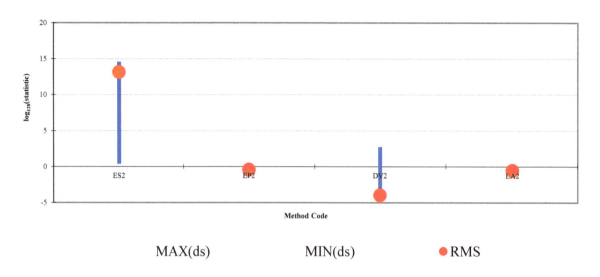

Figure 5

Lastly, in Figure Six we see the wholly different complexion for Range One ($1 \leq z \leq 12$) as the Davis and Lanczos techniques fail completely and various enhanced Stirling forms take over. On the whole the asymptotically-extended Stirling Formula is decisively superior, though Warren Type I is the next best for RMS error and has the advantages of much greater simplicity and a rather narrower defect range.

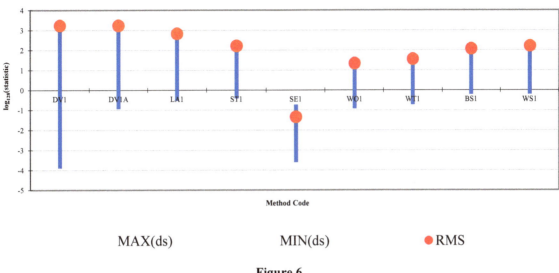

Figure 6

MEF Relations

MEF confers an appreciation of the average performance of a method with respect to establishing significant digits of $\Gamma(z)$.

MEF is listed by method in Appendix B and is illustrated in Figure Seven.

We can see that the Davis Series Method betters eight digits of accuracy in the two lower ranges and returns zero figures of accuracy on Range One. Exponentiated Summation collapses completely in Range Two to return a notional -26.66 equivalent digits: A fact too subtly disclosed by other statistics.

Otherwise you cannot average more than three figure accuracy with anything anywhere except that in Range One the Extended Stirling Formula gives you an MEF of 5.186 and Warren's Type I gives a mean of 3.766 figures.

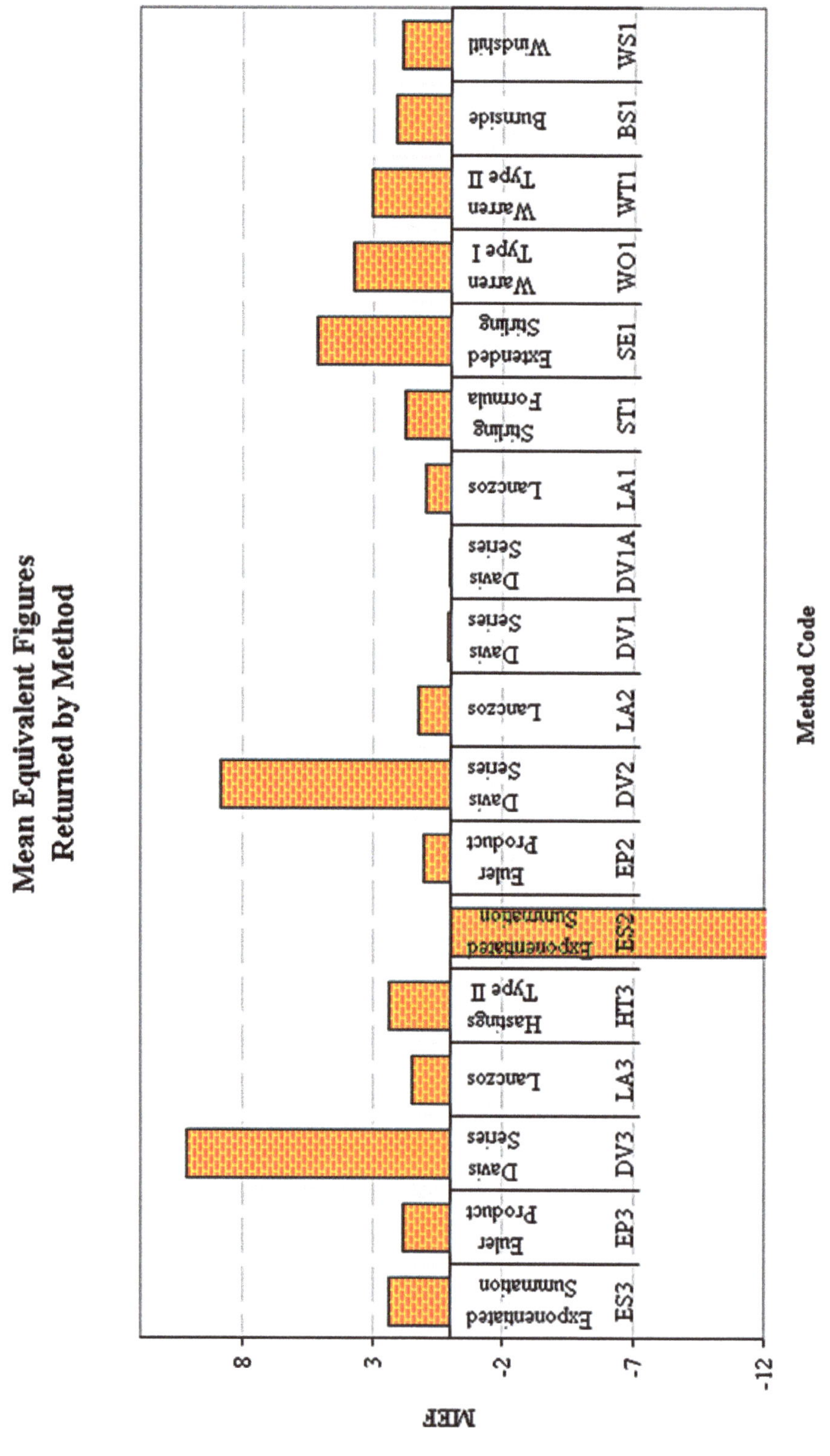
Mean Equivalent Figures
Returned by Method
MEF
8
3
-2
-7
-12
Exponentiated Summation
ES3
Euler Product
EP3
Davis Series
DV3
Lanczos
LA3
Hastings Type II
HT3
Exponentiated Summation
ES2
Euler Product
EP2
Davis Series
DV2
Lanczos
LA2
Davis Series
DV1
Davis Series
DV1A
Lanczos
LA1
Stirling Formula
ST1
Extended Stirling
SE1
Warren Type I
WO1
Warren Type II
WT1
Burnside
BS1
Windshill
WS1
Method Code

Figure 7

CHAPTER FOUR

Some Empirical Approaches to Method Speed

The Trial System

The empirical speed trials were programmed using a high-level programming language. The system employed was a X86-based Mesh Matrix 1.3GHz VPM domestic personal computer fitted with 256MB of RAM and an AMD Athlon Authentic 1.312GHz processor. A math coprocessor was *not* fitted. The total virtual memory on the 37GB hard disk was 2GB, of which 1.67GB were available. This assembly was made sometime prior to May 2001.

The operating system was MicroSoft Windows Millennium Edition 4.90.3000 Build 3000.

I programmed the Complete Gamma Functions solution subroutines and timing structures in Microsoft Visual Basic 6.0(SP4) Learning Edition, a simple object-oriented HLL.

A feature of VB6.0 Learning Edition is that it does not permit you to record *compiled* programs, but does allow you to save, in addition to various source objects, a *consolidated* "made" executable file as a unitary data entity composited of various project objects, necessary and sufficient to the designed application in hand. The VB6.0 Professional and Enterprise editions permit the recording of true machine-code object programs, in p-code or native code. These true assemblies and compilations may be designer-optimised for celerity or for storage-minimisation.

We may expect true compilations to offer much greater celerities than interpreted source, and as we shall see in the next section, "making" a consolidated .exe file conferred no speed advantages: If anything rather the contrary.

Notwithstanding these restrictions, and the general mediocrity of the system itself, it is possible to elicit valuable information about the relative performance of the mathematical structures that interest us.

Repeatability Issues

The timing methods used relied upon the repeated performance of each solution method many thousands of times, in order that reliable average solution times could be established when individual functions elaborated so quickly. Indeed, implied solution times measured in microseconds were commonplace, even for interpreted VB on an old Mesh Matrix.

As with the Accuracy studies of the previous chapter I computed 101 solution points in up to three standard ranges of argument z, but added an outer iterative loop of one to five thousand cycles called Number of Cycles, N_c.

Such a procedure as this raises concerns about both the proportionability and precision of repeat performances as N_c varied.

With regard to proportionability we should hope that five thousand cycles would take five thousand time units; sixty cycles sixty time units; *pro rata* through the range of N_c. That is to say that N_c would be a rectilinear function of time wholly epitomised by a first-order algebraic polynomial.

Number of Iterations	Total Interpreted SPEED Execution Time (seconds)	xy	x^2	y^2	$(x-\mu_x)^2$	$(y-\mu_x)^2$
1	2.47	2.47	1	6.1009	1171188	4409613
2	4.89	9.78	4	23.9121	1169024	4399455
3	7.19	21.57	9	51.6961	1166863	4389812
4	9.44	37.76	16	89.1136	1164703	4380389
5	11.98	59.9	25	143.5204	1162546	4369763
10	23.78	237.8	100	565.4884	1151789	4320569
20	47.4	948	400	2246.76	1130425	4222934
40	95.02	3800.8	1600	9028.8004	1088296	4029486
80	189.28	15142.4	6400	35826.918	1006439	3659943
1000	1929.97	1929970	1000000	3724784.2	6924.617	29724.47
2000	3853.68	7707360	4000000	14850850	840496	3067059
3500	6813.78	23848230	12250000	46427598	5840853	22197310
3500	6765.94	23680790	12250000	45777944	5840853	21748812
5000	9678.47	48392350	25000000	93672782	15341210	57397172
SUM	15165	29433.29	105578960	54508555	204501940	38081610	1.43E+08
MEAN	1083.21429	2102.3779	7541354.3	3893468.2	14607281	2720115	10187289
$(\Sigma)^2$	229977225	866318560	1.115E+16	2.971E+15	4.182E+16	1.45E+15	2.03E+16

n | 14

a | 6.11688403
b | 1.93522279
r | 0.99998917
r^2 | 0.99997834

Table 2

With regard to precision our desire is that the linear regression of N_c upon time would intersect all experimental points snugly, yielding a Correlation Coefficient r very close to unity.

I intentionally attempted to mitigate secondary storage interrupts by programming results' committals outwith timed sections of elaboration, and my observations of disk behaviour suggested that virtual memory was never resorted to: And indeed I never programmed such. We may therefore be assured that all timed transactions elaborated in the solid state.

Table Two summarises the statistics of fourteen timing experiments using the VB6.0 code of SPEED.vbp presented in Appendix C.

The mean Execution ("Mill") Time for a batch of 101 point solutions *solved once* in a z range is 8.05210682 seconds. In general, a first-order polynomial was found to fit the data with the equation:-

$$T = 6.11688403 + 1.93522279 N_c$$

Equation 40

Technically-speaking the intercept is unreliable, the actual one-pass execution time being a mere 2.47 seconds.

More importantly though, we can see that r = 0.99998917 and the Coefficient of Correlation, r^2, therefore 0.99997834. This implies that 99.997834% of the variation of T as N_c is accounted for by structural factors, especially software formations, leaving 0.002166% of the variation to be accounted for by vagaries.

Figure Eight illustrates the almost total linearity of the N_c-time relationship up to $N_c = 80$ and the essential absence of an intercept. That may mean that, because of its semi-compiled state, the VB project requires no preparatory lead-time to execute a set of gamma determinations.

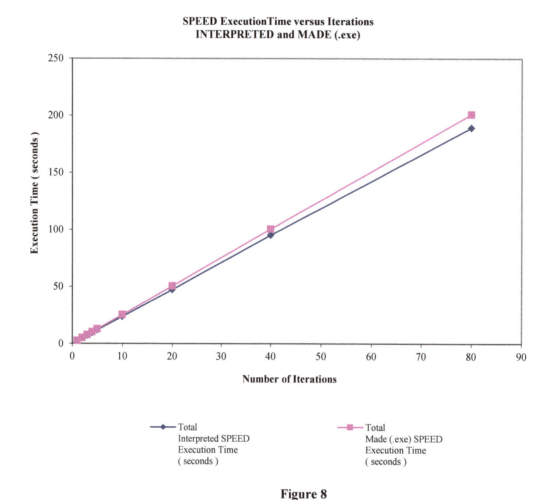

Figure 8

Figure Nine shows the virtually perfect rectilinearity of the complete suite of fourteen experiments and its regression.

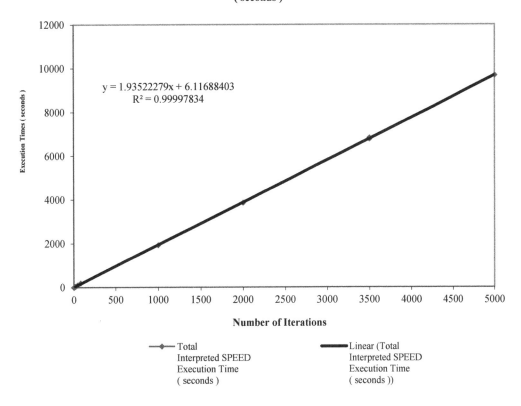

Total Interpreted SPEED Execution Time
(seconds)

$$y = 1.93522279x + 6.11688403$$
$$R^2 = 0.99997834$$

Execution Times (seconds)

Number of Iterations

Total
Interpreted SPEED
Execution Time
(seconds)

Linear (Total
Interpreted SPEED
Execution Time
(seconds))

Figure 9

We may accordingly confidently conclude that both the proportionability and precision aspects of timing experiment repeatability are wholly adequate to our projected assessments of Gamma Function relative speeds.

Timing Tariffs

These empirical timings are based upon the Timing Tariff experimental principles, often used in the treatment of elementary operations like assignment or addition.

As already touched-upon, *absolute* elaboration times are highly implementation-dependent, can be perturbed by interrupts, and of course vary greatly between compiled and interpreted codes.

Relative elaboration times are always determined experimentally for particular systems and thus from a design standpoint have retrospective limitations. They may however inform low-level structural design considerations, and such are the essence of this study.

Clock Dependency

Experimental tariff runs consult some intrinsic system clock function that should obtain real time from the system oscillator circuit. In practice, our program SPEED.bas depends upon the old single-precision MicroSoft intrinsic TIMER which is calibrated in seconds (since last midnight) but accurate only to the second decimal place (centisecond). This coarse structure is the more reason to employ extended highly-iterated procedures to get accurate average times.

TIMER zeros at midnight and care was taken that no program run straddled that datum.

Methodology

In order accurately to time the actual process of interest *and to discount extraneous factors*, even its iterative matrix, we must take certain structural precautions.

We must resolve the Operation ("Racehorse"), or mathematical assembly of interest, from the Carcass, which is everything else that serves and houses it.

Essentially we time $n \times N_c$ applications of the Operation within the setting of its Carcass; time $n \times N_c$ applications of the empty Carcass alone; and then deduct the latter from the former to derive a time that is *due to the action of the Operation alone*.

If we identify Nloop = N_c = Number of Iterations we may program the timing of a composite addition-assignment operation a=b+c with the following pseudocode:-

```
Nloop:=5000
--Carcass Timing
        t1:=Timer()
        for i in 1...Nloop loop
                end loop;
        t2:=Timer()
        tcarcass:=t2-t1
--Operation Timing
        t3:=Timer()
        for i in 1...Nloop loop
                        a:=b+c
                end loop;
        t4:=Timer()
        toperative:=t4-t3
        tdone:=(toperative-tcarcass)/float(Nloop)
```

We can see that tdone is the time to perform one addition-assignment operation, as if that were executable in isolation.

Algebraically, allowing that T is the Operation Execution Time, t_1 is Carcass Start Time, t_2 is Carcass Finish Time, t_3 is Composite (Operation plus Carcass) Start Time and t_4 is Composite Finish Time, then:-

$$T = \frac{(t_4 - t_3) - (t_2 - t_1)}{N_c} = \frac{t_4 - t_3 - t_2 + t_1}{N_c}$$

Equation 41

Digit Frequency

Digit Frequency is a simple rate-of-gain metric, the dividend of Mean Equivalent Figure by Operation Execution Time:-

$$\phi = \frac{MEF}{T}$$

Equation 42

This variable has the dimensions of T^{-1} and is therefore a frequency, the average number of correct output digits formed by a given computational method in each second of elaboration.

This simplest interpretation of Digit Frequency assumes of course a linearity and continuity of performance most uncharacteristic of Complete Gamma Function solution algorithms.

We have seen how Classical solution series are theoretically infinite but not necessarily constant in their rate of precision gain, and in any case shall eventually founder for lack of buffer-space in any realisable machine. On the other hand, Stirling-based techniques will eventually diverge or else are frankly designed only to achieve n-figure estimates.

Notwithstanding this, ϕ is a readily-conceivable, easily-calculable and altogether handy first-order metric of algorithm quality.

The numerical values of ϕ are $\sim 10^5$ to $\sim 10^7$ (Hertz) for our set of Complete Gamma Function solution algorithms.

A Review of Speed Results

Program SPEED.bas and its associated project ancillaries outputted clock results for a range of gamma solutions. Since each range was divided into one hundred equal intervals this meant that 101 *point* solutions were determined in each range.

Therefore, machine-submitted times for the various runs were divided by 101 to determine average *point* values for T.

Statistics were calculated for INTE1000, INTE2000, INTE3500, INTE3500A and INTE5000. These five otherwise identical run jobs had respective N_c iterations of 1000, 2000, 3500, 3500 and 5000.

Appendix D presents the "Accuracy-Speed Ratios" worksheet data that gives the individual method averages as $T \times 10^6$, together with Execution Time Population Mean, μ, Population Standard Deviation, σ and Coefficient of Variation, CV, defined as $100\sigma/\mu$.

Because times are separated by four orders of magnitude, their logarithms to the base 128 are also tabulated for convenience.

Similar tabulation of accuracy statistics are also presented here for comparative convenience, and Digit Frequency averages computed for each method.

Figure Ten shows the means and $\mu\pm\sigma$ spreads of Point Execution Times by method.

Note that the ordinate is calibrated in $\log_{128}(T\times10^6)$.

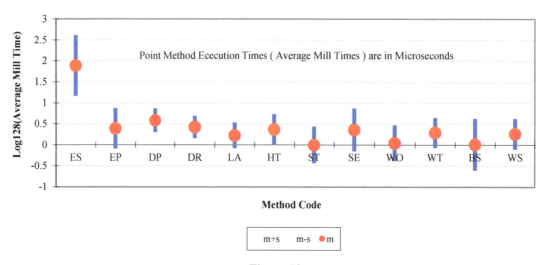

Figure 10

We see immediately the Exponentiated Summation ES is about 550-times more time-expensive than the next most costly method, Davis Series Summation DV, and is the most uneven method in terms of σ spread.

The swiftest methods are Stirling Formula estimators such as the unembellished 1730 Stirling Formula ST, Warren's Type I Formula WO and Burnside's Formula BS.

The Davis Series, expressed as a summed exponentiation DP and as a continued product DR, showed the narrowest dispersion of times across the five N_c experiments.

To assist conception, these Point Mill Times are re-expressed in the logarithmic bar chart of Figure Eleven:-

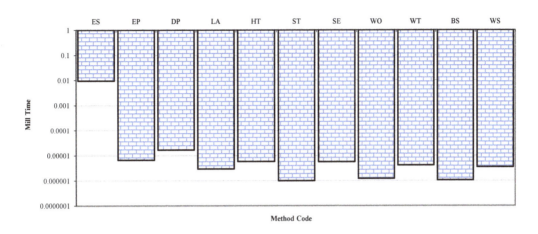

Figure 11

which confirms the speed advantages of the uniterated, closed form formulae ST, WO and BS.

Figure Twelve represents the Method Accuracies in terms of the Mean Equivalent Figure, MEF, per Method.

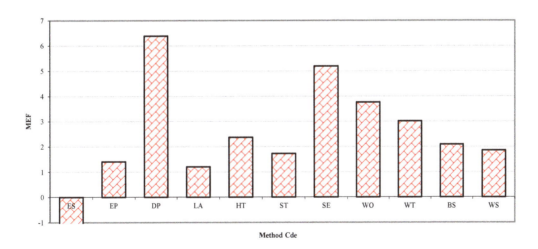

Figure 12

This again emphasises the higher z collapse of Exponentiated Summation, ES, and shows that the Davis Series DP is decisively the most accurate method for all three ranges with better than six-figure accuracy. This is followed by the Extended Stirling Formula, SE, with about five-figures and Warren Type I, WO, with four. The Euler Product, EP, and Lanczos Approximation, LA, are seen to be surprisingly inaccurate.

Finally in this chapter, Figure Thirteen shows the wide dispersion in the quality metric Digit Frequency by Method:-

Digit Frequency by Method

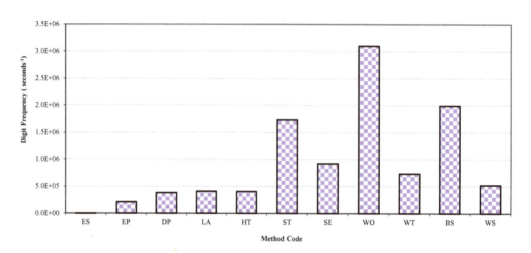

Figure 13

We see that the Euler Product, EP, and the Davis Series, DV, are very inefficient with ϕ less than 500,000. Closed-form formulas are often much more efficient, the best yield rate being given by Warren Type I Formula, WO, at about three million significant figures per second.

CHAPTER FIVE

The Execution Speeds of Component Operations and Functions

To address the design of Complete Gamma Function (CGF) solution algorithms in a more systematic and deliberate way we need, amongst other stratagems, to reduce such algorithms to their elementary mathematical operations and function invocations. These invocations include software-structural devices such as iterations and accesses.

Such analyses will also throw light upon Halstead measurements and other assessments of complexity, affording we hope deeper insights into solution algorithms' comparative utilities and the issues affecting those.

An observer may suppose that if the times taken by entire solution procedures exhibit a linear increment with N_c Numbers of Cycles applied (which is essentially proven in Chapter Four); then barring freak or contrivance their component operators will also exhibit linear execution time versus number-of-applications.

In practice, this proportionable time response is *not* absolute. But time consumption behaviours are *sufficiently* linear over a wide range of N_c to allow us to measure tenable *relative* operation and function average execution times.

As these average execution times are established for each chosen operation or function we build a Timing Tariff: A list of the comparative times for each component operation. Within the constraints of programming practicality this enables the swiftest configuration of operations and operands to be formulated.

Operations and Functions

The mathematical operations and functions we address are naturally those which comprise our various Complete Gamma Function solution procedures, together with some which are likely to suggest themselves for incorporation in future developments.

The eleven selected operators fall into these classes:-

A   **Cardinal Operators**
MULtiply; DIVide; ADD and SUBtract

B   **Exponentiations**
Variable to Arbitrary Power (PWR)
Square of Variable (SQR)

C   **Data Management Operators**
ASSignment; Access of Variable in Brackets (BRT)
One Iteration of a FOR-NEXT Loop (LAC)
Access of a Variable in a One-Dimensional Array (AR1)
Access of a Variable in a Two-Dimensional Array (AR2)

The nine chosen component functions are:-

D   **Napierian and Root Procedures**
Napierian Exponential (EXP)

Napierian Logarithm (LGN); Briggsian Logarithm (L10)
Hyperbolic Sine (SNH); Square Root (SRT)

E   **Reimann's Zeta** (ZET)

F   **Circular Functions**

Arctangent (ATN); Cosine (COS); Sine (SIN)

There are accordingly twenty operators and functions and we need not make any theoretical distinction between such, as we consult only their computational efficacies.

For convenience we list these elements, with their Method Codes, in Table Three:-

1	PWR	B^C
2	SQR	B^2
3	MUL	B*C
4	DIV	B/C
5	ADD	B+C
6	SUB	B-C
7	ASS	B:=C
8	BRT	(B)
9	EXP	EXP(B)
10	LGN	LOGN(C)
11	L10	LOG10(C)
12	ZET	ZETA(B)
13	SNH	SINH(A)
14	SRT	SQRT(B)
15	ATN	ATAN(B)
16	COS	COS(A)
17	SIN	SIN(A)
18	LAC	Loop Access
19	AR1	1D Real Array Access
20	AR2	2D Real Array Access

Table 3

Repeatability Issues

In theory, the old GW-BASIC intrinsic function Timer permits us to measure elaboration times to the nearest centisecond and given sufficient Numbers of Cycles this was good enough to time the solution algorithm composites.

The elementary functions and operators, however, solve too swiftly, even in interpreted Basic, satisfactorily to be treated with Timer.

We must resort to some more precise time estimator, and the MicroSoft Visual Basic library function GetTickCount() was found adequate to our purposes. Whilst Timer returns a Single representing seconds-since-nominal-midnight to two places of decimals, GetTickCount() returns a Long Integer representing the number of milliseconds elapsed from Start. GetTickCount() is therefore ostensibly ten times as precise as Timer.

Notwithstanding that, the absurdity of negative execution times often manifested for the fastest operations such as cardinals, assignments or array lookups. Negative timings persisted even for experiments where N_c was in the hundreds of thousands. Accordingly, negative timings have been discarded in the computation of statistics, including tariff averages.

Seventeen timings experiments were applied with Numbers of Cycles N_c ranging between 10000 and 4000000. In each trial the total amount of time utilised for operation or function computation was about 92.8% of the entire program elaboration time.

Figure Fourteen shows the essential linearity of response aforementioned, which underpins the basic usefulness of a timing tariff that lists average mill times as constants.

TG and TH Series Total Time

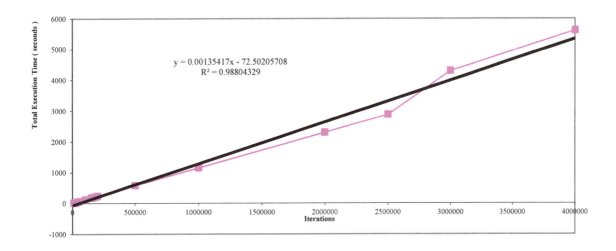

Figure 14

Having discarded all negatives, we should note that not all our precision problems are thus solved, because there is severe dispersion in the absolute timings, especially for the cardinals and data operations, as illustrated by Figure Fifteen:-

TG&TH Series Coefficient of Variation Chart

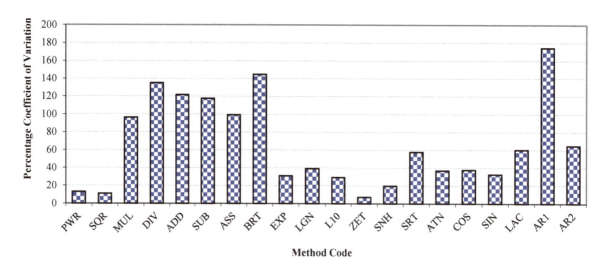

Figure 15

Summary of Findings

Program TARIFF, presented in Appendix E, was applied to the seventeen experiments, and I have assembled its outputs in the worksheet shown in Appendix F.

The Timing Tariff statistical summary is repeated here for convenience, in Table Four:-

		Logn(Mean in Nanoseconds)	Mean (nanoseconds)	Mean	Population Standard Deviation	Coefficient of Variation
1 PWR	B^C	6.1318	460.2431	4.6024E-07	6.6289E-08	14.4029
2 SQR	B^2	6.1491	468.2755	4.6828E-07	4.7215E-08	10.0828
3 MUL	B*C	2.7191	15.1667	1.5167E-08	2.1608E-08	142.4720
4 DIV	B/C	4.5613	95.7060	9.5706E-08	1.3172E-07	137.6274
5 ADD	B+C	3.6466	38.3449	3.8345E-08	5.2345E-08	136.5112
6 SUB	B-C	3.6662	39.1035	3.9104E-08	5.4484E-08	139.3339
7 ASS	B:=C	3.8846	48.6490	4.8649E-08	5.1137E-08	105.1146
8 BRT	(B)	3.4398	31.1806	3.1181E-08	3.1107E-08	99.7631
9 EXP	EXP(B)	6.0517	424.8264	4.2483E-07	1.4982E-07	35.2665
10 LGN	LOGN(C)	4.9070	135.2273	1.3523E-07	6.4940E-08	48.0227
11 L10	LOG10(C)	5.9522	384.5949	3.8459E-07	5.0872E-08	13.2274
12 ZET	ZETA(B)	13.8765	1062908.9931	1.0629E-03	3.7260E-07	0.0351
13 SNH	SINH(A)	6.8791	971.7593	9.7176E-07	3.3795E-08	3.4777
14 SRT	SQRT(B)	3.9141	50.1042	5.0104E-08	3.4043E-08	67.9442
15 ATN	ATAN(B)	5.0619	157.8914	1.5789E-07	5.8514E-08	37.0598
16 COS	COS(A)	4.6784	107.5926	1.0759E-07	4.3667E-08	40.5855
17 SIN	SIN(A)	4.9364	139.2708	1.3927E-07	5.1733E-08	37.1458
18 LAC	Loop Access	5.0502	156.0532	1.5605E-07	1.0757E-07	68.9297
19 AR1	1D Real Array Access	4.6214	101.6389	1.0164E-07	1.3864E-07	136.4014
20 AR2	2D Real Array Access	4.4463	85.3125	8.5313E-08	6.5288E-08	76.5281

Table 4

Figure Sixteen orders the component operators and functions in increasing mean execution time (Napierian logarithm of nanoseconds for graphical clarity).

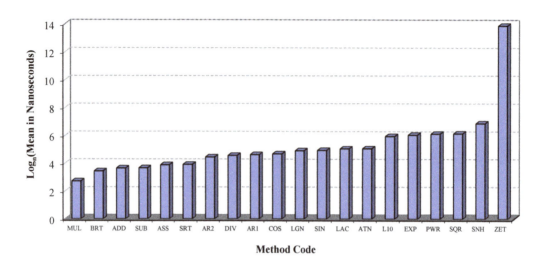

Operations and FunctionsTime Costs

Figure 16

Figure Sixteen omits the high-N_c TH-series timings which do not in principle modify the following observations:-

1   MULtiplication is the swiftest computational operation but BRT Bracket Resolution is quicker than ADDition or SUBtraction and surprisingly all four are quicker than ASSignment.

DIVision is half again as expensive as multiplication.

Consulting the table of absolute averages substantiates the old dictum that it is cheaper to multiply a variable by itself rather than square it: Indeed if you wish the twenty-fifth power it is more time-economical to multiply the variable by itself twenty-four times, or up to x^3 with a loop including an assignment.

2   Interrogating a one-dimensional array (AR1) is almost twice as expensive as a two-dimensional array (AR2), a factor of high potential interest in programming algorithms which generate series intermediates like Davis Coefficients or Stieltjes Numbers.

3   The SQR() Square Root intrinsic is almost eight times swifter than $x^{0.5}$.

4   A Napierian Logarithm (LGN) costs one third a Base-10 one, which latter has no VB intrinsic.

5   Circular function calls are about as time-costly as one another.

6   My simple-minded, and slowly-convergent, method of calculating Zeta by means of its definitional series, is obscenely expensive: A real source of procedural delay.

CHAPTER SIX

Code Metrics

Code metrics attempt to measure the properties of an algorithm or other software system in terms of the syntactic elements in which it is actually or potentially written.

These code metrics may illuminate the programmatic efficiency of the computational process in terms of speed, storage or accuracy; or throw light upon the probable error content of the composition, and therefore its reliability as a realised software segment.

In this chapter we will touch upon these genuses of code metrics:-

(a) **Line Counts**

These are simply the number of statements, identified with "lines" of code, that occur in the given segment.

(b) **Token Metrics**

These are derived from counts of HHL code tokens, definable as operands and operators. They include such as the Halstead Metrics.

(c) **Cyclomatic Complexities**

These are graph theoretic constructs based upon the number and pattern of decisions, and decision branches, in a code assembly. In our context, where Complete Gamma Function solution algorithms elaborate determinately to some approximation, we should note that a pre-specified FOR-NEXT iteration implies a comparison-based terminator decision.

In his book *Software Engineering with Modula-2 and Ada*[20], Richard Weiner accents the implications of software metrics in terms of complexity and its implications for intelligibility on the part of the human programmers charged with maintenance.

He initially reminds us of the cardinal, and general, forms of metric:-

1 **Nominal**

The subjective verbal assessment of a program: "simple", "moderately complex", "complex", "extremely complex".

We are compellingly reminded of Lord Kelvin's aphorism regarding knowledge of a meagre and unsatisfactory nature.

2 **Ordinal**

Classifications based upon rank, which are relative and conserved under normal mathematical transformation.

The virtue of the speed and accuracy measures discussed in prior chapters of this disquisition is that whilst many of the *absolute* measures may be special to interpreted Visual Basic 6.0, the *relative* standings of the various CBF methods should conserve under platform transformations.

3 **Interval Measure**

The spacing between ranked measureands is tenably assessed.

It is likely, but not established, that our CBF methods metrics like Execution Time and Digit Frequency are Interval Measures, conserved under hardware and language transformations.

4   **Ratio Measure**

The highest form of measurement based upon *absolute* and intrinsic properties of the measureand.

A ratio measure is susceptible of measurement in terms of fiducial units of mass, space and time.

At a mathematical level, Execution Time and Digit Frequency are of course dimensional physical measures, but we must remember that any metrics of a realised algorithm are perforce contingent upon the supporting machine architecture.

Weiner reminds us further of the fact that a true ratio scale has a zero and yet how do we cogently define "zero complexity" in a computer program?

If we wish to establish code metrics less rigorously as Interval Measures, then how are we to calibrate our cognitive ruler with "units of complexity"?

Accordingly, only a relativistic Ordinal Measure of software complexity is epistemologically-tenable whatever our *practical* instruments of gage.

Line Measures

When we attempt simplistically to measure algorithm size in terms of "numbers of lines" we encounter, besides the coarseness of scale, immediate ambiguities of definition.

Weiner asks us what a "line of code" is: Is it an executable statement; do declarations, or even comments, qualify as lines of code?

Is it a line of code in the HLL (which HLL?) or is it a line as the object machine instruction or some compilation intermediate?

None of these issues are susceptible of sufficient resolution to make a Line Measure a consistent and reliable industrial tool and accordingly we must seek for finer metrics.

Halstead's Software Science

In his 1977 book *Elements of Software Science*[21], MH Halstead attempted to refine the code metric in terms of counts of the numbers of distinct tokens employed in a code segment, together with counts of the numbers of times those tokens were employed.

For these purposes a token is either a:-

1   **Operand**

This is a numerical constant, or a symbolic name surrogating a numerical constant or variable. It terms of such a definition an operand includes an array element.

2   **Operator**

This could be a mathematical operator such as a cardinal operation or an exponentiation, but in the programming context Halstead extends the definition to include executive keyword primitives such as *If*, *For*, *Next* or *Then*.

Certain ambiguities do of course remain.

In our context of Complete Gamma Function (CGF) solution algorithms it is not clear whether an array access or even a bracket-resolution qualifies as an operation.

And invocation of a function opens a whole new can of worms: Is the function call a single operation or an evolved nest of operands and operators?

In our CGF treatments we will adopt the simple view that array-access and bracket-resolutions or operations and function calls are single operations

In fact we will step a remove even more simple than that. We will treat individual typographical elements of operators as whole operators: Thus (( )) are four operators and the array reference A(2,3) would contain the operands A, 2 and 3 and the operators (, and). By the same token, if you will forgive the pun, EXP is a single operator but EXP(x) contains three operators.

It is notable that in the terms in which it is couched Halstead Theory is general enough to measure any HLL code or even algorism; and yet realised computer code swells to a multiplicity of more elementary operands and operators as it compiles. It is this latter one-to-many resolution of a code token to a series of low-level elementary instructions which is used to justify the typological counting methodology of Halstead enumerations. Notwithstanding all this, it remains unclear how these conventions affect the generality of Halstead metrics in practice.

Taken together, this discrimination between Operators and Operands, and between the Repertoire of Types and the Number of Applications, suggests four countable "independent" variables. These are listed in Table Five.

Number of Distinct OPERATORS **applied** in the algorithm	n_1
Number of Distinct OPERANDS which **appear** in the algorithm	n_2
Total Number of OPERATOR occurrences	N_1
Total Number of OPERAND occurrences	N_2

Table 5

This led Halstead to derive a family of *Predicted* and *Realised* code parameters. His most primitive parameter is the Program Vocabulary, n:-

$$n = n_1 + n_2$$

Equation 43

The Predicted Program Length, N', is given by:-

$$N' = n_1.\log_2 n_1 + n_2.\log_2 n_2$$

Equation 44

whilst the (realised) Program Length, N, is simply:-

$$N = N_1 + N_2$$

Equation 45

Besides Program Length, there is a further first-level derivative of the countable variables, Program Volume.

Realised Program Volume, V, is definable as:-

$$V = N.\log_2 n = N.\log_2\left(n_1 + n_2\right)$$

Equation 46

It is perhaps easiest to access Potential Program Volume, V*, in terms of the Volumes Ratio (or [approximate] Program Level), L', which is given by:-

$$L' = \frac{U}{V} = \frac{2}{n_1} \times \frac{n_2}{N_2}$$

Equation 47

and L'≤1.
Accordingly:-

$$V^* = U = LV$$

Equation 48

Knowledge of L' and V allows us to determine the Program Intelligence Content, I, as:-

$$I = L' \times V$$

Equation 49

that Halstead empirically observed to correlate with total programming and debugging time. Program Intelligence Content, I, remains relatively invariant under translation from one programming language to another.

Finally, the Program Development Effort, E, is given by:-

$$E = \frac{V}{L} \approx \frac{V}{L'}$$

Equation 50

Examples of Halstead Variable Counts

In *Software Engineering*[22], Stephen R Schach offers a small illustration of how we may apply Halstead Theory to variable counting for n_1, n_2, N_1 and N_2.

He gives the HLL code segment:-

```
if(k<2)
{
```

```
        if(k>3)
            x=x+k;
    }
```

in which Operands are coded blue and Operators are coded red.
Schach resolves the Number of District Operators, n_1, as 10, listing them thus:-

if, (, <,), {, >, =, *, ; and }

The Number of Distinct Operands, n_2, is 4:-

k, 2, 3 and x

The Total Number of Operators, N_1, is 13:-

```
    if( < )
    {
        if( > )
            = + ;
    }
```

and the Total Number of Operands, N_2, is 7 elaborated as:-

```
    k  2

        k  3
        x  x   k
```

In terms of our CGF algorithms coded in VB6.0 we might select Exponentiated Summation that I coded in Subroutine METHODxEP as:-

```
DSUM=0#
For KK=2 To 10
  COKK=KK
  DSUM=DSUM+((-1)^KK*ZETA(COKK,2000)*ZZ^KK)/KK
Next KK
GG=1/(ZZ*EXP(ZZ*EMC-DSUM))
```

Treating the function calls as operators we identify N_1 as the count of:-

```
        =
For     =  To
        =
```

$$= \quad +((\ )^\wedge \quad *ZETA(\quad , \quad)* \quad ^\wedge \quad)/$$

Next

$$= /(\quad *EXP(\quad * \quad - \quad))$$

which *appears* to be 31. But we remember that N_1 is the number of *applications* of this population of operators.

Therefore:-

$$N_1 = 1 + \left[(10 - 2 + 1) \times (3 + 1 + 15 + 1) \right] + 10$$
$$= 1 + (9 \times 20) + 10$$
$$= 191$$

Meanwhile, the count of *distinct* operators, n_1, is shown by:-

$$= \text{For To} + (\ ) \ ^\wedge \ * \ \text{ZETA} \ , \ / \ \text{Next EXP} \ -$$

so n_1 is 14.

Turning now to the Operands we may establish the count of *distinct* operands, n_2, from:-

DSUM 0# KK 2 10 COKK -1 2000 ZZ GG 1 EMC

which counts to $n_2 = 12$.

Now these operand citations occur outwith the loop:-

DSUM 0#
GG 1 ZZ ZZ EMC DSUM

a total of eight, whilst these:-

KK 2   10
COKK KK
DSUM DSUM   -1 KK   COKK 2000   ZZ KK KK
KK

are within the loop, which is 15 operands within the loop.

So to compute N_2 we write:-

$$N_2 = 8 + \left[(10 - 2 + 1) \times 15 \right]$$
$$= 8 + 9 \times 15$$
$$= 143$$

Now the foregoing analysis is all very well in a way. But it exemplifies another major ambiguity that bedevils code analysis: The status of iterations.

Of its very nature, an iteration is a stereotype whose *information* inheres in its pattern of progress, whilst for sure we hope and expect that each iterate is different enough to produce a distinct phenotype.

Therefore, token application counts N_x logically include the contents of iterate expansions and may be expected strongly to correlate with empirical measures of speed and accuracy including Execution Time and Digit Frequency.

And yet the accent of Halstead Theory and other code metric analyses is the content of the executive recipe and its implications for human and machine performances and the reliability of the software product.

So when we count Halstead metrics in the context of CGF methods we can only expect a strong correlation with empirical metrics for methods without true iteration (i.e. those modified by a fixed number of corrector terms, if any). This basically covers the Stirling Formula group of methods and also the Lanczos Approximation.

The corollary is that for Classical methods and for other infinitely-iterable forms the Halstead *information* content of the method does not strongly correlate with empirical algorithm performance in terms of its phenotype *outcome*. And the greater the number of iterations, the less the association between genotype and phenotype in our sense of statistical correlation.

In summary, code metrics and empirical metrics treat of different aspects of algorithm quality, and illuminate different desiderata of design.

Given, then, that we will include only the one exemplar of iteration in our counts of token applications, and therefore side-step the semiotic redundancy implied in iteration, we may accordingly revise our N_1 count to 31 and N_2 to 23 in the above Exponentiated Summation analysis.

The Expectation of Token Ratios

In a normal sequence of simple algorism we anticipate that there will be an alternation of operands and operators, commencing and finishing with an operand.

Let "x" stand for any Operand (Variable or Constant) and "+" stand for any Operator or Function. For clarity, we shall again show Operands in blue and Operators or Functions in red.

Then we expect this code pattern:-

$$x+x+x+x+...+x$$

An example of such a pattern is:-

$$E = m * c ^ 2$$

This principle of alternation extends to computer instantiations of algorism, including FORmula TRANslator and all successive coding systems including Visual Basic 6.0

Accordingly, Halstead's N_1 and N_2 are not entirely orthogonal.

For sure, strict alternation is sometimes broken as in the nesting of parentheses or something, but our statistical *expectation* is that N_2 is N_1+1:-

$$E\{N_2\} = N_1 + 1$$

Equation 51

The implication is that the *ratio* of N_1 and N_2 is likely to convey more information about the character of the mathematical process than the actual absolute values themselves.

Therefore, if The Token Count Ratio is R, then the expectation of R is given by:-

$$E\{R\} = \frac{N_2}{N_1 + 1} = 1$$

Equation 52

Accordingly, an "Operand-Rich" algorithm shows R>1, a "Balanced Algorithm" R=1, and an "Operator-Rich" algorithm R<1.

To maximise operational economy we may, at least in theory, seek to maximise R by reducing the operator count to the feasible minimum.

We can illustrate the matter by again referring to our basic Exponentiated Summation algorithm. Remember that Operands are shown in blue and Operators or Functions in red:-

```
DSUM=0#
For KK=2 To 10
  COKK=KK
  DSUM=DSUM+((-1)^KK*ZETA(COKK,2000)*ZZ^KK)/KK
Next KK
GG=1/(ZZ*EXP(ZZ*EMC-DSUM))
```

Once again, our token counts are
$N_1 = 31$
$N_2 = 23$
Therefore R is:-

$$R = \frac{23}{31+1} = \frac{23}{32} = 0.71875$$

So Exponentiated Summation is an expensive, "Operator-Rich" method.

Cyclomatic Complexity

Like Halstead Theory, Cyclomatic Complexity constructs attempt to predict algorithm complexity, and document the complexities of realised algorithmic products.

This is done to compare programs and program modules in terms of human and machine efforts applied to production; module performance; and thirdly reliability issues especially in regard to error predictions.

As with Halstead and empirical metrics, our focus in the context of Complete Gamma Function solution methods is upon machine performance.

The theme of Flow Graph Theory, that discloses Cyclomatic Complexity, is not Operators or Operands, but the populations of Conditionals (decisions) which inhabit an algorithmic module.

Flow Graph Theory is a more or less straight-forward adaptation of Eulerian Graph Theory to computer programming interests.

Consider the Flow Chart of Figure Seventeen.

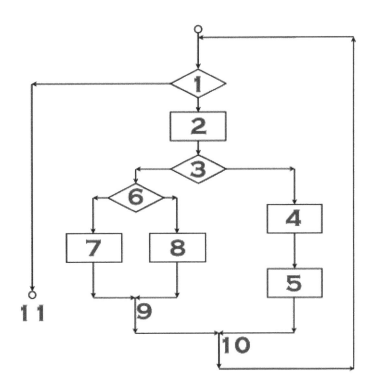

Figure 17

Denoted by the lozenges are three decisions, which we assume to be non-compound, e.g. a=b, a≤b, a≥b or something.

Compound decisions which implicate OR, AND, NOR or NAND would require a slight complication of the graph topology that we need not pursue.

Denoted by the five rectangles, there are additionally five elementary code sequences.

In addition to the Conditionals and Sequences we have labelled selected logical junctions and termini to facilitate analyses.

We shall need to mirror this Flow Chart with an equivalent topological Flow Graph.

Allow that each Node of the Flow Graph is one or more procedural statements. Procedural statements can be conflated to any successor conditional: A Node that contains a Conditional is a Predicate Node.

A Predicate Node spawns two or more Edges: In this case the Predicate Nodes are P1, P2 and P3.

Code branches rejoin at notional Flow Chart junction points which themselves render to Flow Graph Nodes.

In the Flow Graph, circles represent Nodes; arrows are Edges, (where the arrow direction indicates elaboration flow), and edge-bounded areas (including the area outwith) are Regions: In this example regions R1, R2, R3 and R4.

The equivalent Flow Graph is shown in Figure Eighteen:-

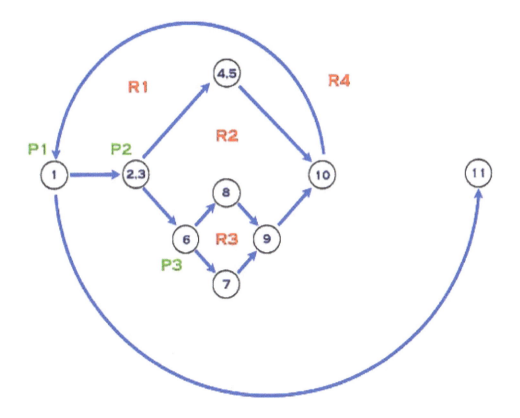

Figure 18

Cyclomatic Complexity is a metric that quantifies a code segment's logical complexity.

Cyclomatic Complexity is a computable integer that equivalates:-

A   The Number of Flow Graph Regions

B   The Number of Independent Basis Paths

C   The Maximum Number of Tests Required

Note that in this context a "test" is supposed to identify with a specific trajectory of elaboration: It need hardly be said that the tests of a mathematical construct should be rather more searching than that.

Independent Paths

An Independent Path introduces at least one new set of processing statements or a new condition. It traverses at least one unused Edge.

The Basis Set for the example graph is:-

        Path 1:   1-11

GAMMA
Gamma Solution
      Page 49 of 214
      13:50 Friday, 05 August 2022

Path 2:   1-2-3-4-5-10-1-11
Path 3:   1-2-3-6-8-9-10-1-11
Path 4:   1-2-3-6-7-9-10-1-11

The Computation of Cyclomatic Complexity

Allow that V(G) is the Cyclomatic Complexity of Graph G; E is the Number of its Edges; N the Number of Nodes; P the Number of Predicate Nodes and R the Number of Regions. Then:-

$$V(G) = E - N + 2 = P + 1 = R$$

Equation 53

In the above example structure E = 11, N = 9, P = 3 and R = 4. Accordingly:-

$$V(G) = 11 - 9 + 2 = 3 + 1 = 4$$

Complete Gamma Function's solution algorithms are cyclomatically very simple, and should be if their designer has done his job properly.

It is of course the case that several unitary and determinate high-level functions may resolve to algorithms which implicate several decisions, represented cyclomatically by predicate nodes. Such implied decisions may pre-empt error conditions or select modified procedures for arguments of different types or sizes.

But not withstanding that we are entitled to take a view, from a high-level perspective, that things like exponential or square-root functions are sequential black boxes that do not complicate the gross structures of CGF algorithms.

Accordingly, Predicate Nodes are adjudged absent in Stirling Formula based CGF estimators where there are no high-level iterations and elaboration passes "straight through" almost certainly invoking exponentials, and often roots or even hyperbolic sines.

A class of estimators, such as the Davis Series polynomial, the Lanczos Approximation and the Stirling-based Warren Type II method employ a set of constants in finite iterative correctors. At a technical level, these have a single predicate node embedded in the FOR-NEXT clause and thus have an augmented Cyclomatic Complexity. In theory, however, you could program such finite methods with no embedded decision.

This predicate node is established in its full degree in the infinitely-iterable Classic methods such as the Euler Product.

In summary, we may tabulate CGF cyclomatic complexities as:-

Method	V(G)
Stirling-based (not WT)	1
Warren Type II	2
Davis Series polynomial summation	2
Lanczos Approximation	2
Classical convergents	2

Table 6

Clearly, CGF methods of higher complexity can be developed including those containing compound conditionalities, e.g. those that terminate after a given number of significant figures resolve.

CHAPTER SEVEN

Some Statistical Theory

In his 1983 text *Software Engineering*[23], Martin Shooman makes several observations regarding the semiotic concordances of computer programs, focusing upon some elementary properties of their rank statistics.

There is an implied analogy between natural and computer languages.

In particular, Shooman shows that Halstead Metrics can be derived from Shannon Information Theory and are also consistent with the Zipf's Law suite of rank statistics, both of which were independently developed during the interwar period of the last century.

Zipf's Law and Token Counts

Zipf's Law relates Size of entity to its Rank Position in an assemblage of similarly-measured things. In particular, Zipf's Law asserts that there is a linear (i.e. first-order algebraic polynomial) relationship between the logarithm of Rank Position and the Size of the measurand.

Allow that n_r is the Number of Tokens of Type (Rank Position) r and that n is the Total Population of Tokens.

Then the Relative Frequency of Occurrence of Token r, f_r, is:-

$$f_r = \frac{n_r}{n}$$

Equation 54

and:-

$$\sum_{r=1}^{t} n_r = n$$

Equation 55

Zipf's (First) Law is accordingly:-

$$Log_n c = Log_n f_r + a Log_n r$$

Equation 56

which can be re-expressed as:-

$$c = f_r . r^a$$

Equation 57

in which c and a are constants: $a \approx 1$.

Shooman reminds us that Zipf's Law is general enough to apply to a good many geographical, economic and linguistic phenomena.

With regard to token counts in computer programs, Shooman and his reviewed workers detected that whilst the linear relationship was fully-tenable for higher ranks, a curvilinear "tail" supervened for low rank values.

This pattern of ranked-series behaviour was allowed for by the Generalised Zipf's Law:-

$$p_r = \frac{n_r}{n} = \frac{C}{(r + A)^B}$$

Equation 58

In Equation 58, p_r is the Probability of Token Type (i.e. Rank Position) r; and A, B and C are descriptive constants.

Analogising computer and natural languages we may say of the descriptive constants that:-

(1)   B>1

The concordance betrays a limited vocabulary.

(2)   B<1

An extensive vocabulary is detected.

(3)   A>0

The Probability of the First Rank Token is Small.

By reference to the English language, the word that follows "the" has a very low probability.

(4)   A<0

The Probability of the First Rank Token is Large.

Again, in the context of the English language, "of" has a following word of very high probability and that word is "the", as in the phrase "of the".

The implications of the Generalised Zipf's Law for the statistical plot geometry are illustrated in Figure Nineteen:-

Generalised Zipf Plot
LOW-RANK NUMBER CONVENTION

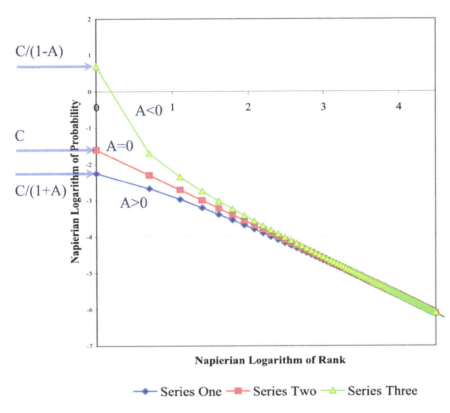

Figure 19

Equation 58 may of course be recast as:-

$$Log_n p_r = Log_n C - BLog_n (r + A)$$

Equation 59

that may assist fitments.

A moment's thought reminds us that Equations 58 and 59 cannot represent a perfectly rectilinear curve *anywhere*: The curve can only vaguely be straight for high values of r.

(And note that it does not matter whether we adopt the convention that big sizes have high ranks, e.g. the largest r = 15, or big sizes have low ranks, i.e. the biggest measureand has r = 1: The underlying mathematics is the same and the log-log plots show indistinguishable geometries).

A further caution is that the Shooman curves of Figure Nineteen are peculiar to very particular combinations of A, B and C. Most conformations of constants yield either crossovers; or absurd probability totals exceeding unity; or both.

Zipf's Law as a Probability Distribution

The foregoing probabilistic interpretation of Zipf's Law is extensible to a full empirical Probability Distribution according to:-

$$\sum_{r=1}^{t} p_r = \sum_{r=1}^{t} f_r = \sum_{r=1}^{t} \frac{n_r}{n} = 1$$

Equation 60

Given that $a \approx 1$ we may substitute n_r/n for f_r in Equation 57 and transpose that to yield:-

$$n_r = \frac{cn}{r}$$

Equation 61

Allowing that for a particular probability distribution, c and n are constants we may expand Equation 61 from the form:-

$$\sum_{r=1}^{t} n_r = cn \sum_{r=1}^{t} \frac{1}{r}$$

Equation 62

Now the RHS summation can be expanded as:-

$$\sum_{r=1}^{t} \frac{1}{r} = \gamma_0 + Log_n t + \frac{1}{2t} - \frac{1}{12t(t+1)} \cdots$$

Equation 63

where γ_0 is the Euler-Mascheroni Constant. Equation 63 converges as $O(t^2)$. Accordingly:-

$$c \approx \frac{1}{\gamma_0 + Log_n t}$$

Equation 64

and:-

$$c = \frac{t}{n}$$

Equation 65

Therefore:-

$$n \approx t\left(\gamma_0 + Log_n t\right)$$

Equation 66

which relates the Number of Token Types, t, to the Total Population of Tokens, n.

Zipf's Second Law

If we adopt the convention that the most populous rank positions have high r and the sparsest have low r, we can define k to be the Number of Occurrences at Low Ranks: i.e. k is a constant n_r where r is large.

Allow that t_k is the Number of Token Types at those Low Rank positions which show $n_r = k$.

Then:-

$$t_k = \frac{cn}{k^2}$$

Equation 67

and:-

$$t = \sum_{k=1}^{n} \frac{cn}{k^2} \approx \sum_{k=1}^{\infty} \frac{cn}{k^2} = \frac{\pi^2}{6} cn$$

Equation 68

Therefore:-

$$n \approx \frac{6}{\pi^2}\left(\gamma_0 + Log_n t\right)t$$

Equation 69

This is essentially the same as Equation 66, but modified by a lesser constant of proportionality, ($6/\pi^2 < 1$).

Bisegmental Zipfian Distributions

Laemmel and Shooman[25] noted that some natural language word distributions displayed two-segment linear log-log relations of the form:-

$$Log_n p_r = \left(\frac{1 + GSWITCH(x_c)}{2}\right)\left(A + BLog_n r\right) + \left(\frac{1 + GSWITCH(x_c)}{2}\right)\left(C + DLog_n r\right)$$

Equation 70

where A, B, C and D are Constants and GSWITCH(x) is a step function that returns -1 when $r < x_c$ and $+1$ where $r \geq x_c$.

The log-log plot of such a system would appear like:-

Segmental Zipf Plot

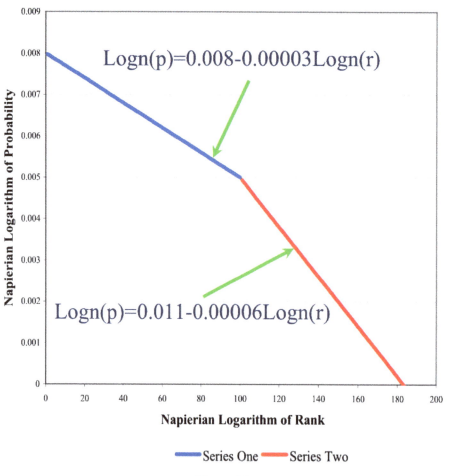

Figure 20

The Identity of Zipfian and Halstead Forms

In his 1983 book, Martin Shooman reminded us that Halstead's Program Length Equation takes the form:-

$$N = n_1.Log_2 n_1 + n_2.Log_2 n_2$$

Now in this context the Number of Token Types $t = n_1 + n_2$ whilst $n \equiv N$. Accordingly we may cast Equation 66 as:-

$$N = (n_1 + n_2)[\gamma_0 + Log_n (n_1 + n_2)]$$

Equation 71

This demonstrates the essential equivalence of Zipf and Halstead concordance treatments, though of course we must remember that both are only statistical estimates, and are not claimed to be appropriate to all situations.

The Shannon Entropy Function

Shannon's Entropy Function is a statistical metric of information content. Shannon's Entropy Function is conventionally denoted as H and summarises the content of an empirical probability series such as that treated by Zipfian rank theory, but it is more general in scope.

Allow that a discrete probability distribution (or a mere array of comparable probabilities) contains i elements, each of which manifest Probability p_j for $1 \leq j \leq i$.

Then:-

$$H = -\sum_{j=1}^{i} p_j . Log_2 p_j = \sum_{j=1}^{i} p_j . Log_2 \frac{1}{p_j}$$

Equation 72

For example, in a language with a Repertoire t of 15 Token Types it is found that in a certain module Type 1 occurs with 50% probability ($p_1 = \frac{1}{2}$) whilst Type 2 and Type 3 each occur at 25% probability ($p_2 = p_3 = \frac{1}{4}$). Meanwhile Types 4 to 15 do not occur (i.e. $p_4 = p_5 = p_6 = ... = p_{15} = 0$).

We may expand Equation 72 as:-

$$H = \frac{1}{2} . Log_2 2 + \frac{1}{4} . Log_2 4 + \frac{1}{4} . Log_2 4 + 0 + ... + 0 = 1.5 \text{ bits}$$

Therefore the said probability distribution may be summarised by 1.5 bits of information.

In situations where the Types are equiprobable, i.e.:-

$$p_j = \frac{1}{i}$$

Equation 73

we may summarise Equation 72 as:-

$$H = Log_2 i$$

Equation 74

On the other hand, in situations where p_js are *not* equiprobable *but* they are governed by Zipf's Law with $f_r \equiv p_j$, then:-

$$H = \frac{N}{Log_n 2} \left[\frac{(Log_n t)^2}{2\left(Log_n t + \frac{7}{12}\right)} + Log_n\left(Log_n t + \frac{7}{12}\right) \right]$$

Equation 75

CHAPTER EIGHT

Some Statistical Observations

This chapter compares selected Complete Gamma Function solution methods in terms of their Halstead Metrics and some other elementary statistics.

The chosen methods are the standard set of eleven plus the Davis Series Summation implemented as a Product Summation (DR).

Halstead Properties

Volumes Ratio

The Halstead Volumes Ratio, L', is defined as:-

$$L' = \frac{2n_2}{n_1 N_2}$$

Equation 76

where n_1 is the Number of Distinct Operators applied in the Algorithm, n_2 is the Number of Distinct Operands which appear in the Algorithm and N_2 is the Total Number of Operand Occurrences.

Figure Twenty-One shows that L' hovers between 0.05 and 0.1 for all Methods except the unembellished Stirling Formula (ST) for which it is 0.238.

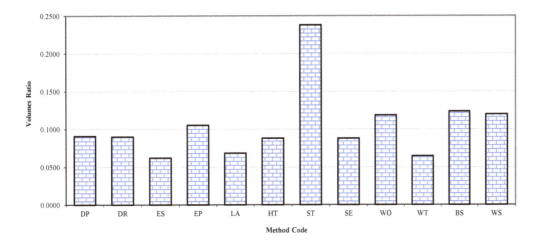

Volumes Ratios by Method

Figure 21

Execution Time Ratios

The Indicated Execution Time, IET, is the sum of the products of Operator Execution Times, OET, multiplied by Number of Operator Applications, NOA:-

$$IET = \sum OET \times NOA$$

Equation 77

It accordingly predicts the Actual Method Execution Time, AET, as determined by experiments.

The Execution Time Ratio, ETR, is defined as:-

$$ETR = \frac{AET}{IET}$$

Equation 78

Clearly, the expectation of ETR is unity. In practice, however, we *anticipate* that ETR will be rather more than unity due to overheads: In practice we *find* it is rather less due to compiler optimisation of the object code.

So for most methods ETR is found in the range 0.6 to 1.0, except Windschitl's Formula (WS) which has an ETR of 1.335, exhibiting significant non-optimisation, possibly due to the presence of square root intrinsic functions (the Sinh function should not make a difference because it is comprised of EXPs which do not adversely affect the several other methods that incorporate EXP). ETR results are illustrated in Figure Twenty-Two.

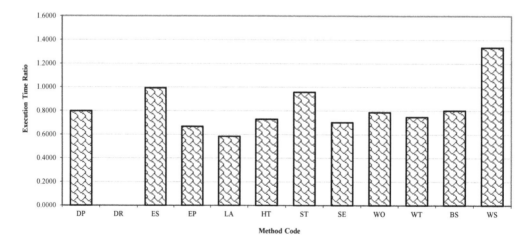

Execution Time Ratios by Method

Figure 22

Token Count Ratios

The Token Count Ratios concerned are those inclusive of iterations, denoted by RI. Accordingly RI counts the actual physical applications of tokens during module elaboration.

Hence the definition:-

$$RI = \frac{A_N}{A_T + 1}$$

Equation 79

where A_N is the Number of Operand Applications and A_T is the Number of Operator Applications.

The expectation of RI is unity.

The RI's for the twelve methods are illustrated in Figure Twenty-Three:-

Token Count Ratio including Iterations by Method

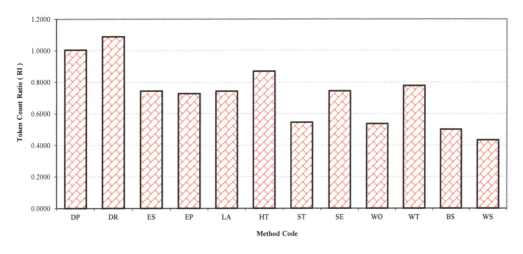

Figure 23

Only the two Davis Series methods (marginally) exceed unity being relatively "operand-rich". The other methods are "operator-rich", more especially Stirling Group methods which show an RI of 0.6±0.18.

There is a discernable positive correlation of RI with formulaic complexity, dramatically shown by ST versus SE; and WO versus WT.

Program Intelligence

Halstead's Program Intelligence Content, I, is given by:-

$$I = L' \times V$$

$$= L' \times \frac{1}{Log_n 2} \times N \times (n_1 + n_2)$$

$$= \frac{L'(N_1 + N_2)}{Log_n 2}(n_1 + n_2)$$

Equation 80

It is sensitive to the amount of *apparent* analytical content in an algorithm and for the CGF methods treated has a mean value near to 116, varying from 64.8 for Burnside's Formula (BS) to 181.5 for Warren's Type II estimator (WT). Other high-content methods are the Lanczos Approximation (LA) with I = 180, and the Extended Stirling Formula (SE) with I = 155.4.

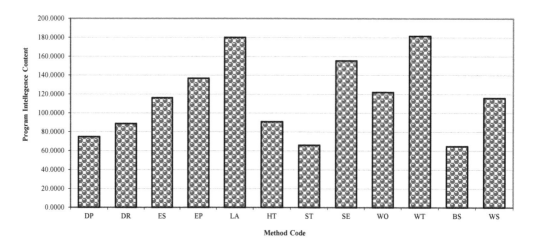

Program Intellegence Content by Method

Figure Twenty-Four illustrates the Intelligence Content:-

Figure 24

Tabular Summary

Table Seven offers six-figure numerical values by method for the metrics discussed above.

Method Code	Volumes Ratio	Execution Times Ratio	Token Count Ratio (RI)	Program Intelligence Content
DP	0.090909	0.797880	1.003448	74.757834
DR	0.090000	0.000000	1.088328	88.812307
ES	0.062112	0.992233	0.744792	116.132470
EP	0.105263	0.668198	0.728395	136.676372
LA	0.068376	0.584237	0.743590	180.028612
HT	0.088235	0.728555	0.869231	90.889788
ST	0.238095	0.957012	0.545455	65.951773
SE	0.087912	0.702245	0.744000	155.367158
WO	0.118519	0.787313	0.535714	122.084060
WT	0.064516	0.748249	0.777778	181.500344
BS	0.123457	0.802253	0.500000	64.832222
WS	0.119658	1.335437	0.433333	116.007478
Mean	0.104754	0.758634	0.726172	116.086702
SD (σ)	0.044971	0.295691	0.191046	39.367672
CV (%)	42.929533	38.976796	26.308623	33.912301

Table 7

Halstead and Empirical Metrics

In order to assess the degree of association between selected Halstead Metrics and the key experimental metrics Actual Execution Time, Mean Equivalent Figure and Digit Frequency, a series of y on x Determination Coefficient calculations were completed.

The Correlation Coefficient, r, assesses the quality of (linear) least-squares correlation between the dependent y and the independent variable x. r varies from zero for utterly no relationship between the variables to ±unity for perfect straight-line coördination. The Determination Coefficient, r^2, gives an enhanced sense of the degree of association, because it directly represents the proportion of variance of y "accounted for" by x. Again $r^2 = 0$ means no relationship at all and $r^2 = \pm 1$ indicates total and exclusive determination of the value of y by the value of x. And $r^2 = 0.38$ means that 38% of the variance of y is set by influence x, whereas 62% is determined by other and perhaps unknown factors to be found.

A merit of a simple measure of association like r^2 is that, with due caution, it can be employed as a selection criterion to reserve or dismiss parameter pairs for future research.

In this study the nominally dependent variables included Intelligence Content, I, and Token Count Ratio, RI.

Also the Simple Token Count Ratio, R, was included defined as:-

$$R = \frac{N_2}{N_1 + 1}$$

Equation 81

and Realised Program Length, N:-

$$N = N_1 + N_2$$

Equation 82

Clearly, R and N are wholly anorthogonal re-arrangements of the same independent variables. We expect them to display identical r^2 with respect to any x.

The actual independent x's are the said empiricals Actual Execution Time, MEF and Digit Frequency.

Table Eight summarises the twelve values of Determination Coefficient.

y	x Actual Execution Time	MEF	Digit Frequency
Intelligence Content	0.0000000630	0.0000269164	0.0370173442
Realised Program Length	0.0353415396	0.0353063499	0.0962605937
Token Count Ratio (RI)	0.0009057871	0.0001211954	0.3858071725
Token Count Ratio (R)	0.0000047089	0.0021188384	0.2898194551

Table Entries are in Determination Coefficient, r^2

Table 8

There is a very striking disparity between $r^2\{R,x\}$ and $r^2\{N,x\}$ in every case that betrays the dependence of r^2 upon the *expressive idiom* of data rather than what the human mind may please itself to regard as the *scientific reality* of the datum.

Accordingly, we should apply the greatest circumspection in both the framing and the review of correlative studies.

Notwithstanding that, I am satisfied that there is no real relationship between Halstead and Empirical metrics, though the $r^2 = 0.38$ determination between RI and Digit Frequency is worthy of further study.

As an empiricist I incline to think that Empirical Metrics are useful and Halstead Metrics not informative so far as CGF methods are concerned.

With regard to the RI versus ϕ relation, linear regression supplies the following equation:-

$$RI = 0.84043652 - 0.00000013\phi$$

Equation 83

for $r^2 = 0.38580717$.

It must be reiterated that this relation is much more tenuous than the eight figures suggest, especially since three of eleven plotted methods out lie at high-ϕ values.

Figure Twenty-Five relates the regression line to the eleven method points.

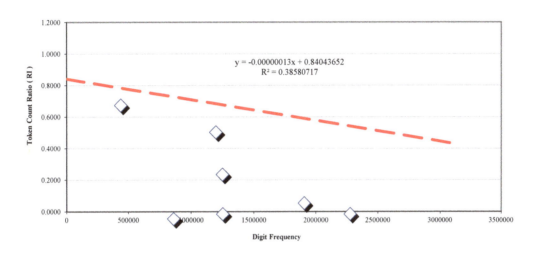

Digit Frequency Versus Token Count Ratio by Method

Figure 25

Zipfian Distributions

Zipf's Law distributions are ranked frequency distributions of discrete quantities. Conventionally, they associate the logarithm of the measureand with rank position serial number, or its logarithm.

Many different kinds of mathematical formula may be fitted to a Zipfian distribution with more or less precision. First-degree algebraic polynomials fitments and power-law fitments are common in the literature but I have shown elsewhere[24] that higher-degree algebraic polynomials are sometimes superior.

The applicability of different formulas to different ranked distributions reflects the type of the underlying statistical distribution that governs the data, e.g. are the data normally-distributed, Poissonian, or whatever.

Otherwise than this the scientific implications of a given Zipfian fit should not be overstated.

Appendix G "Zipfian Plots and Tabulations" presents plots of $Log_n(n)$ versus Rank for our thirteen *applied operations and functions*, together with tabulations.

In this context, n is the sum of all the *Operations or Functions* only of Type O, which are actually referenced in computations. There are twenty classified operations or functions but AR1 (one-dimensional array access), LGN (Radix e Logarithm), L10 (Radix Ten Logarithm), ATN, SIN, COS and SQR (squaring) are never used by any of our twelve CGF methods.

The plot "All Methods Rank versus $\text{Log}_n(n)$" presents the thirteen operations points without interpretation.

The next graph "All Methods Rank versus $\text{Log}_n(f_r)$" shows that data, geometrically identical to that treating $\text{Log}_n(n)$, overlain by four algebraic polynomial regression curves.

As the degree of the fitted polynomial is increased the Determination Coefficient, r^2, increases pseudo-asymptotically in this way:-

Degree	r^2
1	0.8153
2	0.9070
3	0.9576
4	0.9670

Table 9

Theoretically, the twelfth-degree polynomial will be a perfect fit.

I do not seriously suggest that the token content of CGF subroutines is governed by a quartic or any other arbitrary polynomial. What I am attempting to illustrate is that a ranked distribution can be fitted to almost any mathematical structure and that those structures bandied about lack both explanatory and industrial merit.

The plot "General Zipf's Law Model for All Methods" presents the thirteen operators' points in terms of Rank R versus the Empirical Operation Probability, f_r.

The fitted equation is:-

$$f_r = \frac{C}{(R+A)^B} = \frac{3614.955}{(R+14.1483)^{3.605345}}$$

Equation 84

This power-law fitment shows an r^2 of 0.875251, making it slightly better than a straight-line but not as good as a parabola.

We therefore reasonably conclude that so far as elementary CGF algorithms are concerned, the Generalised Zipf's Law does not describe operations' ranked distributions, because it offers no tenable and testable mathematical rule of prediction. It follows that Equation 84 is nugatory.

Lastly in this chapter, we may study the segmental Zipfian plot fitments of the form discussed by Laemmel and Shooman[25].

The graph "Segmental Zipfian Plot for All Methods" shows the fit of two linear least-squares regressions to Rank versus $\text{Log}_n(1/f_n)$. The first line fits the eight lowest-ranked (i.e.

most frequently invoked) operations with $r^2 = 0.945624$. The second fits the six highest-ranked (Rank = 8 is a common point) with a $r^2 = 0.976727$. Accordingly we are able to infer that the operator frequencies fall into two distinct statistical distributions. Why this is so I cannot say.

CHAPTER NINE

The Concept of Profile Ideality

Is there such a thing as a Complete Gamma Function solution algorithm that assembles an ideal mix of component operators and functions?

It all depends upon what we mean by "ideal" of course. We have already seen that there is an inherent tension between numerical accuracy of the output and time economy. Maybe at some level ideality is measured by factors such as Digit Frequency that reconcile both desiderata.

The ideal Complete Gamma Function would accurately estimate $\Gamma(z)$ for any z in the range between zero and infinity. As a matter of mathematical definition solutions *at* zero or infinity are of course impossible. But the CGF methods that exist, even the most robust, are restricted to limited intervals of the real continuum. Classical methods tend to be restricted to $0<z\leq1$; Hastings to $1\leq z\leq2$; and Stirling-based methods to $z>2$.

When we adjudge the efficacy of convergent approximations of transcendental numbers the Order of Convergence, controlled by the growth of dominant terms that may be n, nlogn, n! or whatever, is of key interest in the determination of ϕ, and we have not touched upon this.

CGF methods present very limited cyclomatic complexities and short rosters of operators and operands. Therefore we expect Halstead Metrics to be starved of the statistical feedstocks which underwrite their validity, and whether for that or other reasons Halstead and Zipfian measures bring little to tell of CGF algorithms' efficacies.

In this chapter we shall focus narrowly upon Execution Times and upon Profiles of Operations Execution Times to see whether such assist our quest for the "ideal" CGF algorithm.

The ideal algorithm does not, I suppose, utilise only the very cheapest mathematical operators and functions. I should expect it largely to exercise those quickest components, but to season them with a judiciously-applied admixture of slower and more sophisticated tools.

We can therefore assume that the relative duration of employment of an operation or function shall ideally be *inversely proportional* to its execution time. This is of course only the simplest assumption and a developed econometry may apply a more sophisticated rule.

In order to reduce both the ideal operational profile and those of existing CGF methods to a common basis we will need to scale Reciprocal Execution Times to sum to unity in the same arithmetic way as is conventional when treating probabilities.

When we have done that we can inter-compare Ideality Profiles graphically or as tabulations.

To clarify and summarise such comparisons between real and ideal methods it will, however, be useful to develop certain descriptive statistics.

Let t_j be the Mean Execution Time of the jth Operator or Function, of the Set of n Operators or Functions. We have twenty measured operators from PWR (Power of Real b) to AR2 (2D Real Array Access) but only thirteen operators or functions are actually used by our assessed CGF methods).

Further allow r_j to be the Time Reciprocal for j and f_j to be the Time Fraction for the $j^{th.}$ operator.

Accordingly:-

$$r_j = \frac{1}{t_j}$$

Equation 85

and:-

$$f_j = \frac{r_j}{\sum r_j}$$

Equation 86

where of course:-

$$\sum f_j = 1$$

Equation 87

The array of f_j for j = 1...n, constitutes an Ideal Profile.

With regard to the compared Methods, let g_j be the Actual Fraction of Operations performed by Operator j, and c_j be the Actual Number of the $j^{th.}$ Operator applications during Method i.

Then:-

$$g_j = \frac{1}{c_j} \bigg/ \sum \frac{1}{c_j}$$

Equation 88

The array of g_j for j = 1...n constitutes the Achieved Operative Profile for Method i.

Actual Profiles may be compared to the Ideal Profile operator-by-operator in terms of the statistical Squared Deviation, δ_j^2, defined as:-

$$\delta_j^2 = \left(f_j - g_j\right)^2$$

Equation 89

The departure of a Method from the Ideal may be summarised as the RMS of δ_j^2 defined as the Temporal Ideality, T_i, of Method i:-

$$T_i = \sqrt{\dfrac{\sum\limits_{j=1}^{n} \delta_j^{\,2}}{n}}$$

Equation 90

The less is T_i, the more ideal the CGF solution algorithm (Method).

Computed Idealities

Appendix H "Temporal Ideality Profiles" presents a tabulation of Time Fractions for both the Ideal and the Actual Methods. This table is entitled "Comparative Time Fractions".

The chart "Time Fraction Profiles Chart (Squared Deviations)" plots δ_j^2 for each Method.

We can see that Exponentiated Summation (ES) makes rather too much or too little use of SUBtraction and that several methods make atypical appeals to EXP (Napierian Exponentiation), especially the Extended Stirling Formula (SE).

Windshitl's (WS) is highlighted as the only Sinh (SNH) user.

By far the dominant feature of the plot, however, is the very major peak for Davis Series summation; both DP (Power Summation) and DR (Product Summation): This does not necessarily mean these methods *over*use DIVide, a relatively expensive function: In fact they relatively *under*use it.

Therefore, as always, caution is called for in the interpretation.

Methods can be atypically good as well as atypically bad.

In order to develop a better picture of which operators or functions cause most concern from the operator execution times point of view, and which are the offending operators or functions we need to preserve information about the positive or negative *deviation* from the ideal. This is done simply by recollecting the definition of Deviation δ_j, as:-

$$\delta_j = f_j - g_j$$

Equation 91

The chart "Method Ideality Profiles for Time (Deviations)" discloses that the Davis Series methods DP and DR indeed *under*apply the expensive DIVide operator because they have strong negative troughs around –0.8.

ES (Exponentiated Summation) uses less than the "ideal" number of SUBtractions but uses ZETa functions and we have no direct evidence of what *they* invoke.

Because the chart represents only those methods which use a given operator or function, only seven methods are quantified for EXP (Napierian Exponentiation). They are Lanczos Approximation and the six Stirling-based procedures, all of which modestly ration themselves to much less EXPing than is "permissible" by the Ideal. The Extended Stirling Formula (SE) is especially frugal with its nevertheless essential EXPing as is the historically-related Burnside's Formula (BS).

Warren Type I (WO) is seen to make a balanced appeal to a spread of elementary operators. erring on the side of frugality.

Only Burnside (BS) uses SRT (Square Rooting) but twice is a bit excessive (circa +0.08).

The column chart "Method Departure from Temporal Ideality in terms of RMS(Time Fraction Deviation)" displays T_i for the twelve CGF methods. This departure can of course be a "good thing" or a "bad thing", and the Davis Series methods are markedly deviant in some way whilst Stirling-based methods display a more "ideal" mix of components.

You have probably perceived the similarity between this plot and that for the Token Count Ratio RI.

You may also recall the definition of RI in terms of:-

$$RI = \frac{A_N}{A_T + 1} = \frac{Operand\ Applications}{Operator\ Applications + 1}$$

whose expectation is unity.

The scatter graph headed "RMS(Time Fraction Deviation) versus Token Count Ration RI" correlates T_i (as dependent variable) with RI.

The operators' points y upon x linear regression may be expressed as:-

$$T_i = -0.00022379 + 0.14822231 RI$$

Equation 92

with a determination coefficient $r^2 = 0.45854830$, a moderately good degree of association.

Neglecting an insignificant intercept (that should technically demand a new regression) we will write:-

$$T_i = 0.1482 \times RI$$

Equation 93

CHAPTER TEN

Estimates which use Bernoulli Numbers

Is there such a thing as a Complete Gamma Function solution algorithm that assembles an ideal mix of component operators and functions?

I have posed this question before but I now ask you to ponder it in a different way.

Gamma is definable for any real or complex number yet at zero and infinity Gamma is itself infinite.

In between those extremes Gamma, whatever the character of the feed argument, is a non-terminating transcendental number that can only be estimated: It bears no determinate value, even in principle, whatever the quality of our mathematical methods or mechanical tools. (Integer arguments of course yield factorials which are *apparently* terminated integers).

In this sense the continuum of numbers is infinitely "fine-grained" and our best estimate of $\Gamma(z)$ must perforce fail to *be* $\Gamma(z)$ if only by a few significant figures at an infinite remove.

Nevertheless, we may *estimate* a *particular* $\Gamma(z)$ using several methods, indeed *any* method, of which we have discussed maybe thirteen distinct types. And all of those methods are in principle expressible by many identities.

More than a hundred years ago George Cantor explained how we could thus dissect the continuum of the real and complex numbers to resolve several distinct kinds of infinitude nested within more comprehensive infinities.

Assuredly therefore, having discovered our favorite method of the moment we can confidently explore further in the knowledge that we can supersede our finest effort with a myriad milliard more.

And yet the road to infinity is paved with finite numbers...

The Bernoulli Numbers

The Bernoulli Numbers, or the Bernoulli Constants, are irrational fractions. This means that they are non-terminating fractions that arise as the dividend of integers

A feature of such Irrational Numbers is that their mantissas bear an infinite number of repeating digit cycles, if you will forgive the pleonasm. An example is $1/7 \approx 0.142857142857...$

You possibly recall that a Transcendental Number differs in that *there is no pattern* to the infinite progression of digits.

There are an infinite number of Bernoulli Numbers, and an infinite number of them are infinitely big. All odd Bernoulli Numbers are however zero, whilst the first seven are unity or less. The eighth is seven-sixths and then the absolute values of the Bernoulli Numbers rise dramatically to reach about 6192.123 for the twelfth Bernoulli Number, B_{22}.

An infinite sum of these finite numbers, suitably adjusted by rational functions of the argument z, can be used to estimate $\Gamma(z)$.

Bernoulli Numbers are computable from[26]:-

$$B_n = \sum_{k=0}^{i} \frac{1}{k+1} \cdot \sum_{r=0}^{k} (-1)^r \cdot \binom{k}{r} \cdot r^i$$

Equation 94

where:-

$$\binom{k}{r} = {}_r^k C = \frac{k!}{(k-r)! \, r!}$$

Equation 95

Equation 95 defines a Binomial Coefficient.

Special problems attend both the numerical stability and the computational efficiency of both Equation 94 and 95. Those problems are exacerbated when you attempt to apply 32-bit domestic microcomputers to such purposes.

In fact my system could furnish no useful Bernoulli Number above the sixth (B_{12}), but fortunately a reasonable $\Gamma(z)$ estimate, especially for large z's, is possible using only the first six Bernoulli Numbers.

And I was also fortunate to have several further reliable B_n from Sloane[27].

Table Ten lists the first ten Bernoulli Numbers given from that source:-

Number n	Sloane Numerator A000367	Sloane Denominator A002445	Sloane Value
0	1	1	1.000000000000000
2	1	6	0.166666666666667
4	-1	30	-0.033333333333333
6	1	42	0.023809523809524
8	-1	30	-0.033333333333333
10	5	66	0.075757575757576
12	-691	2730	-0.253113553113553
14	7	6	1.166666666666670
16	-3617	510	-7.092156862745100
18	43867	798	54.971177944862200
20	-174611	330	-529.124242424242000
22	854513	138	6192.123188405800000

Table 10

Binomial Coefficient Computational Issues

The Binomial Coefficient expressed in its classic phrasing as Equation 95 offers large internal redundancies for usual k,r: Both of which are of course integers.

It is wise to re-phrase the Binomial Coefficient as various abridged products both to economise on operations and to ease the management of very large numbers on weak equipment.

It is also prudent to test these forms on your object system to make sure that they do nothing disconcerting over the ranges of n and r that you shall apply.

Using MATHCAD® and also EXCEL® I tested four phrasings of Binomial Coefficient solutions on my aging Mesh Matrix domestic computer. All four expressions agreed implicitly in their numerical outcomes, whether spurious or valid.

Form C^I

This is the classic, fully-redundant expression:-

$$C^I = \frac{k!}{(k-r)!\,r!}$$

Equation 96

We can analyse this and the other three expressions in terms of their elementary component Multiplications, M, and Divisions, D. In such analyses we will neglect any implied implementational loop control operations, but assume that r! is computed *once only* and stored.

Accordingly the Operations Tally, T^I, for expression C^I is:-

$$T^I = M[2k-3]+1D$$

Equation 97

For a selection of k≤24 and r≤19, C^I was entirely accurate up to at least i=20.

Form C^{II}

The identity C^{II} is given by:-

$$C^{II} = \left(\prod_{i=r+1}^{k} i\right) r! \Bigg/ \left(\prod_{i=r+1}^{k-r} i\right)(r!)^2$$

Equation 98

Its tally is:-

$$T^{II} = M[2k-3r+1]+1D$$

Equation 99

guaranteeing better performance than C^I whenever r≥2.

It breaks down, however, in a very dramatic way at i≥13, its computed coefficients rapidly "dying-away" to zero at i=10.

Form C^{III}

C^{III}, our third expression for a Binomial Coefficient, is given by:-

$$C^{III} = \left(\prod_{i=k-r+1}^{k} i \right) r! \Big/ (r!)^2$$

Equation 100

The operations tally is:-

$$T^{III} = M[2r - 2] + 1D$$

Equation 101

that guarantees superior performance to C^{I} for all non-trivial choices of k and r.

C^{III} is virtually as stable as C^{I}: In MATHCAD only it differs by 10^{-16} at i=18 and is identical elsewhere.

Form C^{IV}

A simple cancellation of r! in C^{III} yields:-

$$C^{IV} = \left(\prod_{i=k-r+1}^{k} i \right) \Big/ r!$$

Equation 102

whose tally is:-

$$T^{IV} = M[2r - 3] + 1D$$

Equation 103

that guarantees the best performance of the four identities studied.

C^{IV} is equivalent to C^{I} for the twenty studied values. Nevertheless, we should remember that this leaves more than $24! \approx 6.2 \times 10^{23}$ combinations untested!

Notwithstanding that, I recommend form C^{IV} wherever we need to compute a Binomial Coefficient in the absence of a swift and reliable intrinsic.

Cost Profiles

Allowing that my system of interpreted VB6.0 on a Mesh Matrix yielded the temporal cost of Multiplication M as 15.1667 nanoseconds and that of Division D as 95.7060 nanoseconds it was feasible to develop spot cost profiles for i=1...20 at various k and r.

Firstly, form C^I exhibited a constant indicated cost of 778.2075 nanoseconds.

Form C^{II} exhibited a linear *decrease* with i to a negative (!) value at i=20: But remember that the C^{II} process breaks down entirely long before that point is reached.

Forms C^{III} and C^{IV} show close and parallel cost *increases* from 65 or 50 at i=1 to 641 and 626 at i=20.

On the whole, C^{IV} costs 43.48% as much as C^I with a mean of 338.3732 nanoseconds as opposed to 778.2075 nanoseconds. This reinforces our preference for identity C^{IV} in the practical computation of Binomial Coefficients.

The Factor rⁿ/r!

Redundancies intrinsic to rⁿ/r! may repay elimination. The tally of this expression as it stands is:-

$$T\left(\frac{r^n}{r!}\right) = M[n+r-2]+1D$$

Equation 104

Only one identity for rⁿ/r! was examined in this study:-

$$V = r^{n-r}\prod_{j=1}^{r-1}\frac{r}{r-j}$$

Equation 105

Form V was numerically identical to rⁿ/r! for r=1...22 at n=22 and for r=1...22 with n=1...4.

The tally for Form V is given by:-

$$T^V = 1P + 1S + (r-2)[M+D+S]$$

Equation 106

Given that P=460.2431 nanoseconds and S=39.1035 for my system it is easy to show that V is more costly than rⁿ/r! for r>2, and progressively more so as both time costs rise linearly with r.

Accordingly, I counsel that identity V is discountenanced.

A Modified Computational Equation

We are now in a position to propose a stable and more economical form of Equation 94 for the computation of Bernoulli Number values:-

$$B_n = \sum_{k=0}^{i} \frac{1}{k+1} \cdot \sum_{r=0}^{k} (-1)^r \cdot \left[\frac{\prod_{r=k-r+1}^{k} i}{r!} \right] \cdot r^i$$

Equation 107

It should be emphasised that this form is more than twice as cheap, but ultimately no more fecund than the combinatorial form of Equation 94.

A Bernoulli Gamma Estimator

Now that we have acquired a sufficient supply of Bernoulli Constants, and adequately understood relevant limitations of our abilities to manage them, we are in a position to use them in the actual estimation of the Complete Gamma Function.

Many sources, including Abramowitz and Stegun[28] present this:-

$$Log_n \Gamma(z) \approx \left(z - \frac{1}{2} \right) \cdot Log_n z - z + \frac{1}{2} Log_n 2\pi + \sum_{m=1}^{\infty} \frac{B_{2m}}{2m(2m-1)z^{2m-1}}$$

Equation 108

where B_{2m} is the appropriate (even) Bernoulli Number for Term m.

We shall call the form of Equation 108 the "Bernoulli Sum Asymptotic Formula: Log Sum Form" and award it the Method Code AS.

Taking antilogarithms we may write its analog:-

$$\Gamma(z) \approx \frac{\sqrt{2\pi} \cdot z^{z-\frac{1}{2}} \cdot Exp\left(\sum_{m=1}^{\infty} \frac{B_{2m}}{2m(2m-1)z^{2m-1}} \right)}{e^z}$$

Equation 109

This form is the "Bernoulli Sum Asymptotic Formula: Product Form" and has Method Code AM.

The eventual computational forms of these formulas will be simplifications: For a start, we clearly cannot have an infinite sum.

Noting that the first ten B_{2m} do not exceed | 530 | and that the summative series converges as B/z^{2m-1} we find from tests that for any z>2 we can restrict m≤10 for yielded accuracies of the $\Gamma(z)$ estimate that better absolute defects of 10^{-8}. But below z=2 this estimator is wholly useless.

We may also note that the term factors:-

$$c_m = \frac{B_{2m}}{2m(2m-1)}$$

Equation 110

are constants for any given m. Therefore of course all c_m for m=1...10 can be computed once and stored.

Furthermore:-

$$c_s = \frac{1}{2} Log_n(2\pi)$$

Equation 111

and its antilogarithm:-

$$c_r = \sqrt{2\pi}$$

Equation 112

are of course general constants.

Table Eleven lists the first ten c_m Composite Correctors. Their "sag" to an absolute minimum at m=4 exemplifies the relatively high rate of corrective accuracy gain at the lowest iterations of the Bernoulli summand. Unless z is large it is vain to expect significant corrections by higher m terms.

m	$B_{2m}/2m(2m-1)$
1	0.083333333333333300
2	-0.002777777777777780
3	0.000793650793650794
4	-0.000595238095238095
5	0.000841750841750842
6	-0.001917526917526920
7	0.006410256410256410
8	-0.029550653594771200
9	0.179644372368831000
10	-1.392432216905900000

Table 11

An implication of Equation 110 is that whereas true classical convergent methods such as EP or ES estimate $\Gamma(z)$ with O(n), Methods AS and AM provide O(m²) convergence. We must however remember that these Bernoulli methods describe an *asymptote* similar in principle to that of the Extended Stirling Formula Method SE. This means that an m-term is reached at which the exponential growth of the numerator (i.e. B_{2n}) compensates the power decline of the denominator, and we gain no further precision, or even lose that which we already have, as we add new increments to the summand.

In terms of our numerical system this limits us to an ultimate precision of a finite number of significant figures achieved when n is nine or ten (far from the infinity that the theoretical form suggests!). In fact the choice of n is critical as it is with the Stirling approximation. (MATHCAD yields up ten-figure accuracy at z = 3.55 and fifteen significant figures at z = 14).

Thus we can frame a new practical computing equation for Method AS as:-

$$Log_n\Gamma(z) \approx \left(z - \frac{1}{2}\right)Log_n z - z + c_s + \sum_{m=1}^{10}\frac{c_m}{z^{2m-1}}$$

Equation 113

and for the form of Method AM as:-

$$\Gamma(z) = \frac{c_r z^{z-\frac{1}{2}} \cdot Exp\left[\sum_{m=1}^{10}\frac{c_m}{z^{2m-1}}\right]}{e^z}$$

Equation 114

Bernoulli Methods Performances

$\Gamma(z)$ was computed in the range z=3, 3.11, ...14 which of course provides 100 equal intervals and 101 computed points.

Table Twelve presents a report of the Cycle Execution Times ("Mill Times") for twelve timing experiments on both AM and AS for N_c between 1 and 20000. The experiments for N_c=1,100 and 1000 were discountenanced due to the unreliability of their timing results.

Table Thirteen summarises the key empirical outcomes for the two Bernoulli methods.

It is clear that there is little to chose between the AM and AS methods.

Mean Equivalent Figures (MEF) are virtually identical as may be expected, and their Token Count Ratios RI do not significantly differ.

Surprisingly, the AM method is 6.5% more expensive than AS despite the replacement of sums of logarithms with products and this fact reflects in the lower Digit Frequency of AM.

Figure Twenty-Six illustrates the Token Count Ratio RI versus Digit Frequency ϕ relations for all thirteen treated methods. Plotted points are annotated with Method Codes and algorithm classes discriminated with colored isotype infills. Blue denotes Classical Group methods and red Stirling-based methods. The Lanczos Approximation is green and the Hastings Polynomial is purple.

The plot highlights the averageness of the Bernoulli Asymptotic methods in terms of gain-structure position and especially their statistical proximity to the other asymptotic method the Extended Stirling Formula SE.

AM and AS are balanced, accurate and reliable when the quality of the Bernoulli Constants is properly controlled, but like the Stirling-based methods they only come into their own above the z "critical region" 0<z≤2.

Timings in **Seconds per** *Range*

	Method AM			Method AS		
	Job	Number of Cycles	Execution Time (per *cycle*)	Job	Number of Cycles	Execution Time (per *cycle*)
Discounted	AM1	1	0	AS1	1	0
Discounted	AM100	100	1.10E-03	AS100	100	6.00E-04
Discounted	AM1000	1000	8.20E-04	AS1000	1000	7.10E-04
	AM2500	2500	8.76E-04	AS2500	2500	7.64E-04
	AM5000	5000	8.36E-04	AS5000	5000	7.80E-04
	AM7500	7500	8.43E-04	AS7500	7500	7.84E-04
	AM10000A	10000	8.41E-04	AS10000A	10000	9.12E-04
	AM10000B	10000	1.02E-03	AS10000B	10000	9.45E-04
	AM15000	15000	8.38E-04	AS15000	15000	7.79E-04
	AM20000A	20000	8.40E-04	AS20000A	20000	7.81E-04
	AM20000B	20000	8.43E-04	AS20000B	20000	7.80E-04
	AM20000C	20000	1.02E-03	AS20000C	20000	9.47E-04
Mean		12222.22222	8.85E-04		12222.22222	8.30E-04
Population Standard Deviation		6394.924685	7.43115E-05		6394.924685	7.46302E-05
Percentage Coefficient of Variation		52.32211106	8.400817428		52.32211106	8.989382956
Points per Range		101	101		101	101
Mean Point Mill Time (seconds)			8.7582E-06			8.2198E-06
MEF			9.7456E+00			9.7456E+00
Digit Frequency			1.1127E+06			1.1856E+06

Table 12

	AM	AS
Mean Point Mill Time (seconds)	8.758159E-06	8.219839E-06
MEF	9.7456023370	9.7456019267
Digit Frequency	1112745.518	1185619.614
Token Count Ration RI	0.7393617021	0.7540106952

Table 13

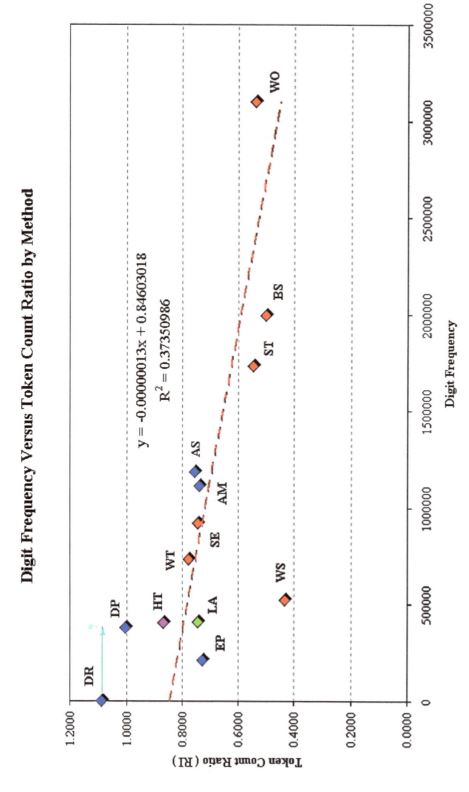

Digit Frequency Versus Token Count Ratio by Method

$y = -0.0000013x + 0.84603018$

$R^2 = 0.37350986$

Token Count Ratio (R1)

Digit Frequency

CHAPTER ELEVEN

Conclusions

CGF Methods Exercised

Method Code	Method Description
AM	Bernoulli Sum Asymptotic Formula: Product Form
AS	Bernoulli Sum Asymptotic Formula: Log Sum Form
BS	Burnside Formula
DR	Davis Series Summation (Products Summation)
DV	Davis Series Solution (Powers Summation)
EP	Euler Infinite Product
ES	Exponentiated Summation
HT	Hastings Type II (Polynomial Approximation)
LA	Lanczos Approximation
SE	Extended Stirling Formula
ST	Stirling Formula
WO	Warren Method Type I
WS	Windschitl Formula
WT	Warren Method Type II

Table 14

(1) System Capabilities

The hardware and software resources employed were found adequate to the elaboration of the said Complete Gamma Function solution algorithms. In particular the platform afforded up to fifteen digits of numerical precision, (conserved in the case of VB programming by the use of double-precision real throughout), and centisecond resolution with Timer or millisecond with GetTickCount(). Timer was used in the Method timing experiments and GetTickCount() in the tariff experiments.

EXCEL® statistical intrinsic functions were adequate to resolve the general discriminations asserted with regard to metric utilities, in so far as such utilities apply to the simple gamma function algorithms discussed.

Without recourse to special tools, the platform described is inadequate to the resolution of Bernoulli Numbers in general, or to certain configurations of other combinatorial products. It is, however, good enough to furnish combinatorially-based $\Gamma(z)$ estimates to about nine figures of accuracy when z>3.

Due care must be taken with regard to the phrasing of estimative algorism and of course its code realisations. Different expressions of identities affect both computational economy and numerical stability, and certain constructions lead to unexpected method failures.

Furthermore, particular CGF methods are only useful for particular argument ranges, and even respected CGF resources such as Davis Series summations or the Lanczos Approximation can exhibit "blind spots" or more extensive intervals of unsatisfactory accuracy.

(2) Empirical Metrics

For details of these important measures of algorithm performance please review Chapter Three and Chapter Four.

The key metric of accuracy is Mean Equivalent Figure, MEF, the average number of leading "correct" digits agreed by the CGF Method concerned and my adopted Fiducial Method.

The formula required is:-

$$MEF \approx -\mu \left[Log_{10} \left| \frac{\Gamma_{test} - \Gamma_{fid}}{\Gamma_{fid}} \right| \right]$$

Equation 115

where μ is the Arithmetic Mean, Γ_{test} a Method Γ Estimate and Γ_{fid} a Fiducial Γ Estimate at corresponding Argument z.

The Mean Method Execution (or "Mill") Time per *point* is separately computed as T, based upon VB6.0 program runs.

This datum enabled definition of the key algorithm implementation efficiency metric Digit Frequency, ϕ, defined as:-

$$\phi = \frac{MEF}{T}$$

Equation 116

Mean Equivalent Figure and Execution Time varied signally between Methods providing powerful discriminants.

These findings are summarised in Table Fifteen.

Method Code	Point Execution Time (ns)	Mean Equivalent Figure MEF	Digit Frequency (Hertz)
DP	16828.10	6.39357575	379934.50
DR			
ES	9505133.24	-12.14742346	-1277.99
EP	6616.83	1.40734932	212692.33
LA	2958.51	1.20424670	407044.98
HT	5829.99	2.36929633	406398.30
ST	997.45	1.72868896	1733101.38
SE	5644.41	5.18622799	918825.04
WO	1214.71	3.76607475	3100389.90
WT	4102.40	3.01234534	734287.74
BS	1049.50	2.09117971	1992539.16
WS	3535.79	1.85529088	524718.24
AM	8758.16	9.74560234	1112745.52
AS	8219.84	9.74560193	1185619.61

Table 15

MEF is excellent at betraying catastrophic method collapses that surprisingly might be overlooked. Such computational failures may not actually generate fatal errors, and are often hidden in other statistics. VALID MEFs are of course positive and collapses are flagged by large negative MEFs.

(3) Relative Errors Analysis

The Root Mean Square (RMS) of the Solutions Difference is:-

$$RMS_\delta = \sqrt{\frac{\left(\Gamma_{test} - \Gamma_{fid}\right)^2}{n}}$$

Equation 117

where n is in this context the Number of Solution Points per Range (i.e. 101). The Percentage Specific Defect, d%$_s$, is:-

$$d\%_s = 100\left(\frac{\Gamma_{test} - \Gamma_{fid}}{\Gamma_{fid}}\right)$$

Equation 118

and for a given solutions range of 101 points the Percentage Specific Defect spreads across a range of its own definable as Max($d\%_s$)-Min($d\%_s$).

RMS$_\delta$ almost always lies sensibly at either Max($d\%_s$) or Min($d\%_s$) for any Method-Solution Range experimental run.

As a measure of spread Max($d\%_s$)-Min($d\%_s$) characterises a Method's *precision* in a given range of Argument z. On the other hand RMS$_\delta$ characterises overall *accuracy*.

When RMS$_\delta$ is high the Method is *improving* its accuracy with increasing z: When RMS$_\delta$ is low the Method is *deteriorating* in accuracy with increasing z.

With the exception of the Lanczos Approximation and the Hasting Polynomial most methods coarsen as they approach the "critical region" 1≤z≤2 from the direction of zero, and refine as they depart from z=2 toward infinity. This statement is of course highly qualified by the fact that specific methods are avowedly optimal only in specific z ranges, or even fail outwith their natural province.

A fuller discussion is provided in Chapter Three and selected results tabulated in Appendix B.

(4) Time Issues

System Time Response to algorithm aggregate repetition was shown to be rectilinear (r^2=0.99997834) and unaffected by the non-compilation of applications software (i.e. Visual Basic Project components). Details are given in Chapter Four.

This certifies the essential repeatability of Method and Component computational timing experiments.

The ETR (Execution Time Ratio) is the proportion of program run time (exclusive of transput operations) that was utilised in actual Gamma routines rather than solid-state systems transactions.

For most Methods ETR lay between 70% and 100% with a mean of 92.8%, indicating a degree of code optimisation of the CGF algorithms over their component operations.

(5) A Timing Tariff

In order to furnish lower-level durational data in regard to the component operators and functions invoked by existing or potential CGF algorithms, VB6 experiments were performed to resolve an Operations and Functions Timing Tariff.

Table Sixteen gives a digest of these Tariff statistics.

Operation or Function Code	Operation or Function Description	Mean Execution Time (ns)	Population Standard Deviation	Coefficient of Variation
PWR	B^C	460.2431	6.6289E-08	14.4029
SQR	B^2	468.2755	4.7215E-08	10.0828
MUL	B*C	15.1667	2.1608E-08	142.4720
DIV	B/C	95.7060	1.3172E-07	137.6274
ADD	B+C	38.3449	5.2345E-08	136.5112
SUB	B-C	39.1035	5.4484E-08	139.3339
ASS	B:=C	48.6490	5.1137E-08	105.1146
BRT	(B)	31.1806	3.1107E-08	99.7631
EXP	EXP(B)	424.8264	1.4982E-07	35.2665
LGN	LOGN(C)	135.2273	6.4940E-08	48.0227
L10	LOG10(C)	384.5949	5.0872E-08	13.2274
ZET	ZETA(B)	1062908.9931	3.7260E-07	0.0351
SNH	SINH(A)	971.7593	3.3795E-08	3.4777
SRT	SQRT(B)	50.1042	3.4043E-08	67.9442
ATN	ATAN(B)	157.8914	5.8514E-08	37.0598
COS	COS(A)	107.5926	4.3667E-08	40.5855
SIN	SIN(A)	139.2708	5.1733E-08	37.1458
LAC	Loop Access	156.0532	1.0757E-07	68.9297
AR1	1D Real Array Access	101.6389	1.3864E-07	136.4014
AR2	2D Real Array Access	85.3125	6.5288E-08	76.5281

Table 16

A timing tariff can be used to assess the overall proportion of the program effort expended upon actual Gamma solutions as opposed to system management, and to throw light upon the degree of software optimisation achieved in given functional codings. On the whole, about 92.8% of elaboration time was expended in actual solutions.

The greatest utility of such empirical tariffs is however for the analysis of existing internal strengths and weaknesses of Method codings; and to inform the choice and frequency of use of components in the design of optimal solutions algorithms.

The tariff determined above offers some surprises but confirms other old maxims of programmers' lore.

Whilst absolute times are platform-specific, the relative standing of functional components (and entire CGF methods) is reasonably robust under hardware and software changes: Which does not of course absolve us from taking a new tariff whenever extensive system changes have been implemented.

(6) Halstead Metrics

In the context of CGF estimation, many ambiguities and contradictions attend our efforts at token enumeration. Some of these are aired in Chapter Six.

Notwithstanding that, we can afford to declare that there is no correlation between classic Halstead Metrics and empirical coded-algorithm performance measures.

The Halstead, and other token-related, metrics treated in this project are listed in Table Seventeen.

Metric Description	Variable Name
Number of Lines	LL
Number of Distinct OPERATORS **applied** in the algorithm	n_1
Number of Distinct OPERANDS which **appear** in the algorithm	n_2
Total Number of OPERATOR occurances	N_1
Total Number of OPERAND occurances	N_2
OPERATOR applications	A_T
OPERAND applications	A_N
Program Vocabulary	n
Predicted Program Length	N'
Realised Program Length	N
Predicted Program Volume	V'
Realised Program Volume	V
Potential Minimum Volume	U
Volumes Ratio	L'
Program Intelligence Content	I
Approximate Program Development Effort	E
Token Count Ratio (simple:- High Values Operand-rich)	R
Token Count Ratio (with iterations:- High Values Operand-rich)	RI

Table 17

The token counts attaching to these variables are tabulated in Table Eighteen.

It is known that Halstead metrical theory subsists upon the semiotic stochastics that attaches to language composition (syntactic structure) and it is almost certainly true that our simple Complete Gamma Function procedures, however coded, contain too few tokens too infrequently applied, and too little cyclomatic complexity, statistically to support Halstead constructs.

	LL	n_1	n_2	N_1	N_2	A_T	A_N	n	N'	N	V'	V	U	L'	I	E	R	RI
DP	5	11	8	14	16	289	291	19	62.1	30	1701.0	822.3	154.6	0.0909	74.8	9045.7	1.066667	1.003448
DR	7	10	9	16	20	316	345	19	61.7	36	1692.6	986.8	152.3	0.0900	88.8	10964.5	1.176471	1.088328
ES	6	14	10	31	23	191	143	24	86.5	54	2995.8	1869.7	186.1	0.0621	116.1	30102.7	0.718750	0.744792
EP	5	10	10	26	19	161	118	20	66.4	45	1917.0	1298.4	201.8	0.1053	136.7	12335.0	0.703704	0.728395
LA	5	13	12	46	27	116	87	25	91.1	73	3286.6	2632.9	224.7	0.0684	180.0	38506.4	0.574468	0.743590
HT	5	12	9	17	17	129	113	21	71.5	34	2167.7	1030.1	191.3	0.0882	90.9	11674.3	0.944444	0.869231
ST	1	7	5	10	6	10	6	12	31.3	16	541.2	277.0	128.9	0.2381	66.0	1163.4	0.545455	0.545455
SE	5	13	12	28	21	124	93	25	91.1	49	3286.6	1767.3	288.9	0.0879	155.4	20103.1	0.724138	0.744000
WO	1	9	8	27	15	27	15	17	52.5	42	1288.3	1030.1	152.7	0.1185	122.1	8691.3	0.535714	0.535714
WT	5	13	13	44	31	89	70	26	96.2	75	3608.9	2813.3	232.8	0.0645	181.5	43605.5	0.688889	0.777778
BS	1	9	5	17	9	17	9	14	40.1	26	810.7	525.1	100.1	0.1235	64.8	4253.6	0.500000	0.500000
WS	1	9	7	29	13	29	13	16	48.2	42	1112.2	969.5	133.1	0.1197	116.0	8102.2	0.433333	0.433333
AM	5	13	11	34	22	187	139	24	86.2	56	2983.2	1939.0	229.5	0.0769	149.2	25206.8	0.628571	0.739362
AS	6	14	11	33	24	186	141	25	91.4	57	3295.0	2055.8	215.7	0.0655	134.6	31398.3	0.705882	0.754011
DSINH	1	6	4	11	4	11	4	10	23.5	15	339.2	216.4	113.1	0.3333	72.1	649.2	0.333333	0.333333
ZETA	4	7	5	9	11	16001	18002	12	31.3	20	541.2	346.2	70.3	0.1299	45.0	2666.1	1.100000	1.124984

Table 18

(7) Token Count Ratios

Whilst actual Halstead Metrics do not apply to CGF algorithms certain other token-derived metrics display weak to moderate correlation with experimental results.

One of these is the (Iterate-Inclusive) Token Count Ratio RI given by:-

$$RI = \frac{A_N}{A_T + 1}$$

Equation 119

where A_N is the Number of Operand Applications actually elaborated in a solution, and A_T the Number of Operator (or Function) Applications.

The expectation of RI is unity.

RI displays a manifest positive correlation with formulaic complexity, though it awards its highest values to algebraic polynomials (Davis methods and Hastings) which a mathematician may find eccentric of it.

RI also shows a reasonable positive least-squares correlation with Digit Frequency, ϕ:-

$$RI = 0.84043652 - 0.00000013\phi$$

Equation 120

with $r^2 = 0.38580717$.

Excepting the two Davis Series methods, RI<1 for all other Methods showing that CGF algorithms are "Operator-Rich", Stirling-based estimators tending to be more so than Classical infinite series techniques.

This gives rise to the apparent contradiction that the more rich in operations an algorithm is, the less its formulaic complexity.

R is the analog of RI for *one iteration only plus non-iterative algorithm elements*. R correlates more weakly than RI. It is only these two Token Count Ratios that correlate with T, MEF and ϕ.

Now the correlations of Execution Time and Mean Equivalent Figure with either ratio are virtually zero in terms of linear least-squares r^2. Therefore R versus T and R versus MEF correlations are occluded ("hidden") and only develop when T and MEF are composited in the ratio ϕ. Thus the need for extreme caution in the interpretation of statistical correlations is re-emphasised. Correlation depends as much upon the *expressive idiom* of data as upon its real or imaginary scientific integrity.

More about this is written in Chapter Five.

(8) Zipf's Law Models

Logarithmic plots of Inverse Token Probability versus Rank superficially describe the bimodal fit geometries described by Shooman: That is to say a rectilinear segment for frequent tokens superseded by a curvilinear form.

It is notable that the General Zipf Law is algebraically incapable of representing such a relation, and only provides a vague fit when provided with special arbitrary parameters.

Indeed it was described in Chapter Eight how other arbitrary curves, especially algebraic polynomials, could provide statistically-superior "explanations" of the p_r versus Rank "relationship".

In terms of the determination coefficient r^2 much better correlations could be achieved by modelling the p_r versus Rank trace as two separate and discontinuous rectilinear segments. (An interpretation also reviewed by Shooman). This leads me to think that in this regard the CGF methods fall into two separate but similar statistical distributions, (*not* bimodality).

(9) Component Profiling

It is possible, from knowledge of the relative execution times of the component operations and functions that might be invoked by coded CGF algorithms, to formulate a typical or "ideal" solutions composition with which planned or existent Methods may be compared.

This procedure can disclose internal "hot-spots" where instabilities or diseconomies may lurk, as well as providing a positive adjunct to mathematical phrase composition strategies.

The convenient approach is to weight operation times by some inverse loading and scale their sums to unity, plotting the Method's deviational profiles as line graphs against an ideal abscissa.

If we load by simple reciprocation of the execution times (as I did in this study) then the required Deviation δ for Component Operation or Function j is:-

$$\delta_j = f_j - g_j = \frac{\dfrac{1}{t_j}}{\sum \dfrac{1}{t_j}} - \frac{\dfrac{1}{c_j}}{\sum \dfrac{1}{c_j}}$$

Equation 121

where t_j and c_j are respectively Method and Ideal Actual Executed Operation Counts, and f_j and g_j respectively Method and Ideal Operation Preferabilities scaled to unity.

A summary deviation can be computed as $T_i = RMS(\delta_j)$ where i labels the Method.

A full explanation may be found in Chapter Nine.

The profile analysis highlights the unusual economy achieved by Davis Series summations in regard to the underinvokation of relatively expensive division operators. Stirling-based procedures always use one or more Napierian exponentiations (EXPs), but are always frugal with respect to the ideal "permissible" utilisation of that function, whereas the Burnside Formula's use of two square-rootings exceeds the normative limit.

The virtue of component profiling is of course that it provides another tool for the low-level optimisation of code phrasings.

A moderately-good linear correlation was detected between the iterations-inclusive Token Count Ratio RI and T_i. Determination r^2 was 0.45854830 for a first-order polynomial that may be simplified to:-

$$T_i = 0.1482RI$$

Equation 122

(10) Bernoulli Asymptotic Methods

Two CGF methods were run based upon the well-known Bernoulli Sum approximation for the Napierian logarithm $\text{Log}_n(\Gamma(z))$.

Whilst this concept suffers from the theoretical incapacity of any asymptotic formula to meet the target, it nevertheless quickly converges to good estimates with a Digit Frequency better than 1.1 MHz for an MEF of 9.7456.

Surprisingly the sum-of-logarithms form Method AS was found to be quicker than the antilogarithmic construction Method AM.

Bernoulli Numbers are employed and the processing of these presents operational stability limitations on 32-bit microcomputers. Nevertheless, given adequately high z and the first ten Bernoulli Constants as double-precision reals we can reliably achieve the accuracy quoted above.

CHAPTER TWELVE

Some Recommendations for Further Research

Numerical Interpolations

Some of the Methods treated in this thesis are actual fitted polynomials (DV, DR and HT) or polynomially-extended derived estimators (WT).

Clearly, such approximating functions are limited with regard to range and precision in principle, though a 26-term behemoth like the Davis Series summation is very useful over a wide interval.

There may be scope for the refinement of short-range estimates using standard numerical interpolation tools where a sufficient number of accurate local fiducial points can be computed.

You can of course employ Lagrangian schemes wherever you can fit an adequate algebraic polynomial, and if you can derive a differential of your estimation formula several other approaches may suggest themselves.

When $\Gamma(z)$ approximates a quadratic, such as in the "critical zone" $1 \leq z \leq 2$, you may essay some modified form of Newton's Method, whilst for $z > 2$ you may prefer to assume (or test the appropriateness of) a geometric progression and apply Aitken's Process.

The problem with all such interpolations is of course that they approximate approximations and bring their own baggage of problems with regard to range, stability, precision and economy.

But this is not to say that they have no rôle in Complete Gamma Function estimation.

AGM Methods

The Arithmetic-Geometric Mean (AGM or simply AG) is quadratically convergent and in certain circumstances can be used to solve elliptic integrals that enable analytic expressions to be derived for particular $\Gamma(z)$ when z is a small rational fraction[29].

Whether this method may be generalised to non-rational z, or to $z \geq 1$, remains an open question.

Central to the argument is a mathematical object called the Complete Elliptic Integral of The First Kind:-

$$K(k) = \int_0^{\frac{\pi}{2}} \frac{dt}{\left(1 - k^2 Sin^2 t\right)^{\frac{1}{2}}} = \int_0^1 \frac{dx}{\left[\left(1 - x^2\right)\left(1 - k^2 x^2\right)\right]^{\frac{1}{2}}}$$

Equation 123

Equation 123 has no general analytic solution but its Modulus k , $0 \leq k < 1$. and its Complementary Modulus, k', where:-

$$k^2 + k'^2 = 1$$

Equation 124

both lead to interesting results.

By recourse to the Arithmetic-Geometric Mean (which he also discovered) Gauss could evolve the *numerical* solution:-

$$K(k) = \frac{\pi}{2 \times AGM(1, k')} = \frac{\pi}{2 \times AGM(a, b)}$$

Equation 125

in a simple and rapid manner.

This is possible because you develop the AGM using a successive-substitution iteration of:-

(1) $\quad a_{n+1} = (a_n + b_n)/2$ **Equation 126**

(2) $\quad b_{n+1} = \sqrt{a_n b_n}$ **Equation 127**

Now allowing that N is a positive integer then:-

$$AGM(1, k'(N)) = AGM[N] = 2 \times K[N] \times \Sigma[N]$$

Equation 128

where:-

$$\Sigma[N] = \alpha(N) - \sqrt{N} \sum_{n=0}^{N} 2^{n-1}\left(a_n^2 - b_n^2\right)$$

Equation 129

and $\alpha(N)$ is a simple algebraic constant involving square roots of integers.

The upshot of these procedures is that you can represent $\Gamma(n/24)$ in terms of AGM, $\Sigma[N]$ and simple fractional powers of 2 and 3 such as:-

$$\Gamma\left(\frac{5}{6}\right) = 2^{\frac{5}{18}} 3^{-\frac{1}{3}} AGM[3]\left\{2^{-1}\Sigma[3]\right\}^{-\frac{1}{6}}$$

Equation 130

Therefore a fine-grained field of "exact" $\Gamma(z)$ in the potentially-difficult range $0 < z \leq 1$ is calculable.

Though these objects look formidable at first sight a careful review confirms that they are easily programmable.

An obvious implication is that you may expect to resolve accurate and stable estimates for $\Gamma(1/n)$ at high values of integer n, thus providing interpolable points in a region where traditional CGF methods are likely to fail.

Borwein and Zucker assert that the overall process exhibits an $O[\text{Log}_n(n)]$ order of convergence.

Better Zeta Values

Simple schemes for the computation of Reimann's Zeta Function, including its definitional reciprocal series, converge at $O(k)$.

$\zeta(k)$ is implicated in several $\Gamma(z)$ estimation formulae and you possibly recall its essential rôle in the Exponentiated Summation Method (ES) that threw in its towel so dramatically when z reached unity. And that even when ζ had had the benefit of two thousand iterations.

To salvage ES for larger z you would need to obtain more accurate zeta functions at the requisite integer arguments k and if you intend to enjoy $O(n!)$ convergence for zeta by means of (e.g.) Equation 17 that will engage the acquisition of a reliable set of Steiltjes Numbers.

An alternative avenue that sidesteps *direct knowledge* of Steiltjes values is defined for the complex argument s=σ+it[30]:-

$$\zeta(s) = \frac{1}{\left(1-2^{1-s}\right)}\left[\sum_{k=1}^{n}\frac{(-1)^{k-1}}{k^s} + \frac{1}{2^n}\sum_{k=n+1}^{2n}\frac{(-1)^{k-1}e_{k-n}}{k^s}\right] + \gamma_n(s)$$

Equation 131

where:-

$$e_k = \sum_{j=k}^{n}\binom{n}{j}$$

Equation 132

and if σ>0:-

$$\left|\gamma_n(s)\right| = \frac{1}{8^n}\cdot\frac{\left(1+|t/\sigma|\right)e^{|t|\pi/2}}{\left|1-2^{1-s}\right|}$$

Equation 133

Clearly, is you want useful zetas at low integral values of s this form will require further development.

The above approach is convergent whenever $\sigma = \Re(s) > -(n-1)$.

Gourdon and Sebah additionally offer the following even faster approach which is valid when s≥½ as of course it would invariably be in CGF computation.

It computes $\zeta(s+it)$ to d decimal digits of accuracy where:-

$$n \approx 1.3d + 0.9|t|$$

Equation 134

$$\zeta(s) = \frac{1}{d_0\left(1 - s^{1-s}\right)}\left[\sum_{k=1}^{n}\frac{(-1)^{k-1}.d_k}{k^s}\right] + \gamma_n(s)$$

Equation 135

where:-

$$d_k = n\sum_{j=k}^{n}\frac{(n+j-1)!\,4^j}{(n-j)!\,(2j)!}$$

Equation 136

and:-

$$|\gamma_n(s)| = \frac{1}{(3+\sqrt{8})^n} \cdot \frac{(1+2|t|)e^{|t|\pi/2}}{|1 - 2^{1-s}|}$$

Equation 137

Both forms are of course singular at s=1, where in any case the estimates of $\zeta(1)$ would be numerically unprofitable.

If you want to confine yourself to real Gammas only you can make some obvious simplifications to the above formulae.

Further doubt remains about the precision of these sophisticated fast algorithms in the low integral s regions of interest to us.

Notwithstanding that Borwein, Bradley and Crandall[31] assert that "…. the functional equation for the Reimann Zeta Function and duplication formula for the gamma function allow one to compute Γ as efficiently as ζ".

Laplace Transforms

Because, explicitly or otherwise, Gamma can be defined in terms of indefinite integrals involving the Napierian base, it may be possible to describe them with a Laplacian of form[31]:-

$$\int_0^\infty f\left(\frac{x}{a}\right)e^{-\mu x}.dx = \sum_{n=2}^{\infty}q_n\big(\zeta(n)-1\big)$$

Equation 138

where the q_n are explicit and rational:-

$$q_n = \sum_{k=0}^{n-2} \binom{n-1}{k} \frac{f_k}{a^{k+1}}$$

Equation 139

Borwein, Bradley and Crandall therefore claim an "intriguing possibility that one may effect numerical integration for some Laplace-transform integrands by way of appropriate rational ζ-series".

Clearly, much depends upon the choice of integrand and the convergence characteristics of q_n.

These authors offer a number of highly-convergent algorithms for $\zeta(n)$ at low values of integer n.

The only explicit CGF estimator that Borwein, Bradley and Crandall offer for further research is:-

$$\left| \Gamma(z) - N^z \sum_{k=0}^{6N} \frac{(-1)^k N^k}{k!(k+z)} \right| \leq 2Ne^{-N}$$

Equation 140

though this is eminently worthy of further analysis and experiment (Reference 31 BBC Page 37).

Interesting in the context of the Borwein and Zucker's previous discussion of Gamma for rational values are the following formulas anent quantum energy levels that appear on Pages 57 and 58 of Reference 31 BBC.

$$z(s) = \sum_n \frac{1}{E_n^{\ s}}$$

Equation 141

and:-

$$z(1) = \left(\frac{2}{(m+2)^2} \right)^{\frac{m}{m+2}} \frac{\Gamma^2\left(\frac{2}{m+2} \right) \Gamma\left(\frac{3}{m+2} \right)}{\Gamma\left(\frac{4}{m+2} \right) \Gamma\left(\frac{m+1}{m+2} \right)} \left(1 + Secant \frac{2\pi}{m+2} \right)$$

Equation 142

where {E_0, E_1, ... } are Eigenvalues of a Specified Quantum System.

BIBLIOGRAPHY AND REFERENCES

Series Number	Detail	Topic
1	"Handbook of Mathematics" IN Bronshtein and KA Semendyayev Translated by KA Hirsch Van Nostrand Reinhold Company of New York Verlag Harri Deutsch of Frankfurt-am-Mein: NAUKA of Moscow December 1978: ISBN 0-442-21171-6: 973pp: Page 91	Factorial Binomial Coefficient
2	http://www.public.asu.edu/~pythagor/euler.htm	Eulerian Functions
3	http://www.weizmann.ac.il/feinburg/courses/ 2002-2000/syllabus2002220.html (university curriculum schedule)	
4	http://www.itl.nist.gov/div898/handbook/eda/ section4/eda4291.htm	
5	"The History of Mathematics" Carl D Boyer John Wiley of New York 1968 LCCCN 68-16506 717pp Page 465	
6	"An Enhanced Estimator for Factorials and for The Gamma Function" James R Warren 9 June 1998 http://www.jamesrwarren.com	
7	"Some Further Enhancement of a Gamma Function Estimator" James R Warren 7 February 1999 http://www.jamesrwarren.com	
8	"Calculators and the Gamma Function" http://www.rskey.org/gamma.htm	Stirling's Formula Burnside's Formula Lanczos Method
9	"Gamma Function" http://mathworld.wolfram.com/GammaFunction.html	Asymptotic Estimator Algorithm
10	"Gamma" Wolfram Research http://functions.wolfram.com/GammaBetaErf/Gamma/06/02	Asymptotic Estimator Algorithm

11   "Handbook of Mathematical Functions"   Davis Series

Edited by Milton Abramowitz and Irene A Stegun
Dover of New York 1965
ISBN 0-486-61272-4
1046pp
Chapter 6 Gamma Function and Related Functions
Page 255

Series Number	Detail	Topic
12	"Numerical Recipes in C" William H Press, Saul A Teukolsky, William T Vetterling, Brian P Flannery Second Edition: Cambridge University Press 1992 quoted in:- "Calculators and the Gamma Function" http://ww.rskey.org/gamma.htm	Lanczos Method
13	Unpublished Email Robert H Windschitl quoted in:- "Calculators and the Gamma Function" http://ww.rskey.org/gamma.htm	Windschitl's Method
14	"Tables of Higher Mathematical Functions" HT Davis Principia Press of Broomington, Indiana 1935	The Davis Series
15	"CRC Concise Encyclopedia of Mathematics" Second Edition Eric W Weisstein Chapman and Hall/CRC December 2002 ISBN 1584883472 3252pp	
16	"The Euler Constant: g" Anonymous 14 April 2004 6pp http://numbers.computation.free.fr/Constants/Gamma/gamma.html	Euler-Mascheroni Constant
17	"Multiple Precision Gamma Function and Related Functions" David M Smith Loyola Marymount University 11pp http://myweb.lmu.edu/dmsmith/FMLIB2.pdf	Bernoulli Numbers
18	"Computing the Incomplete Gamma Function to Arbitrary Precision" Serge Winitzki 14 December 2003 Department of (Theoretical) Physics Ludwig-Maximilians University, Munich, Germany http://www.theorie.physik.uni-muenchen.de/ ~serge/papers/Winitzki S. 2003	Precision Techniques Comparisons
19	"Approximations for Digital Computers" C Hastings Jnr Princeton University Press 1955 Princeton, New Jersey	

Series Number	Detail	Topic
20	"Software Engineering with Modula-2 and Ada" Richard Weiner John Wiley of New York 1984 ISBN 0-471-89014-6 Pages 280-287	Software Metrics Halstead Metric Formulae Cyclomatic Complexity
21	"Elements of Software Science" Maurice H Halstead Elsevier North Holland of New York 1977 ASIN 0444002154 127 pp	Halstead Metric Theory
22	"Software Engineering" Second Edition 1993 Stephen R Schach Richard D Irwin Incorporated of Burr Ridge, Illinois ISBN 0-256-12998-3 Pages 394-397	Token Counting
23	"Software Engineering" Martin L Shooman International Student Edition 1983 McGraw-Hill Incorporated of Singapore ISBN 0-07-057021-3 683 pp Pages 154-203	Zipf's Law Shannon Entropy
24	"Some Studies of the Historical Time Series of British and European Lead Mine Production" James R Warren 8 March 2002 http://www.jamesrwarren.com	Higher-Degree Zipf Law Fitments
25	"Statistical (Natural) Language Theory and Computer Program Complexity" Arthur Laemmel and Martin Shooman Polytechnic Institute of New York Report Number POLY-EE/EP-76-020 15 August 1977	Bisegmental Zipf Law Fitments
26	"Bernoulli Number" Eric W Weisstein From Mathworld - A Wolfram Web Resource http://mathworld.wolfram.com/BernoulliNumber.html	Bernoulli Numbers Computations

Series Number	Detail	Topic

27	"On-Line Encyclopedia of Integer Sequences"
	A000367
	Numerators of Bernoulli Numbers B_2n
	http://www.research.att.com/cgi-bin/access.cgi/
	  as/njas/sequences/eisA.cgi?Anum=A000367
	A002445
	Denominators of Bernoulli Numbers B_2n
	http://www.research.att.com/cgi-bin/access.cgi/
	  as/njas/sequences/eisA.cgi?Anum=A002445
	American Telephone and Telegraph Corporation
	Listed on 12 October 2004
		Bernoulli Numbers
		Values

28	"Handbook of Mathematical Functions" Edited by Milton Abramowitz and Irene A Stegun Dover of New York 1965 ISBN 0-486-61272-4 1046 pp Page 257 Eqn. 6.1.40	Bernoulli Asymptotic Log Sum Method
29	"Elliptic Integral Evaluation of the Gamma function at rational values of small denominator" JM Borwein and IJ Zucker IMA Journal of Numerical Analysis 30 July 1990 1992 V12 pp519-526	
30	"Numerical Evaluation of the Reimann Zeta Function" Xavier Gourdon and Pascal Sabah 10 pp Page Three (Propositions 1 and 2) http://numbers.computation.free.fr/Constants/constants.html	
31	"Computational Strategies for the Reimann Zeta Function" Jonathan M Borwein, David M Bradley, Richard E Crandall Centre for Experimental and Constructive Mathematics Simon Frazer University, Burnaby, British Columbia Report Number: CECM-99-118 30 October 1999: 38 pp	

APPENDIX A

Equation Constants

General Constants

Constant Name	Symbol	Value
The Ludolphine Constant	π	3.141592653589790
The Napierian Base	e	2.718281828459050
Stirling's Root Factor	$(2\pi)^{0.5}$	2.506628274631000
The Euler-Mascheroni Constant	γ	0.577215664901533

Stirling Coefficients

Bernoulli Coefficients   Term Factors

i	a_i	b_i	a_i/b_i	a_i	b_i	a_i/b_i	
1	-1	2	-0.500000	1	1	1.000000000000000	
2	1	6	0.166667	1	12	0.083333333333333	
3	0	1	0.000000	1	288	0.003472222222222	
4	-1	30	-0.033333	-139	51840	-0.002681327160494	
5	0	1	0.000000	-571	2488320	-0.000229472093621	
6	1	42	0.023810	163879	209018880	0.000784039221720	
7				5246819	75246796800	0.000069728137584	Unsafe
8	-1	30	-0.033333	-534703531	902961561600	-0.000592166437354	Unsafe

Warren Coefficients

Cubic Polynomial Coefficients for the Type 2 (W1) Form

	Lower z	Upper z
Range A:	1	2
Range B:	1	12

Coefficient Degree	Range A	Range B
0	0.999798048389	0.999985402531
1	0.000707514046	0.001659904663
2	0.006136240729	0.007674764874
3	0.001339556824	0.002127685900

Lanczos Coefficients

a_0	1.0000000	b_0	1.000000000
a_1	-0.5748646	b_1	-0.577191652
a_2	0.9512363	b_2	0.988205891
a_3	-0.6998588	b_3	-0.897056937
a_4	0.4245549	b_4	0.918206857
a_5	-0.1010678	b_5	-0.756704078
		b_6	0.482199394
		b_7	-0.193527818
		b_8	0.035868343

Hastings Coefficients

i	p_i
0	1.000000000190020000
1	76.180091729471500000
2	-86.505320329416800000
3	24.014098240830900000
4	-1.231739572450160000
5	0.001208650973866180
6	-5.395239384953000000

The Davis Series

k	c_k
1	1.0000000000000000
2	0.5772156649015330
3	-0.6558780715202540
4	-0.0420026350340952
5	0.1665386113822920
6	-0.0421977345555443
7	-0.0096219715278770
8	0.0072189432466630
9	-0.0011651675918591
10	-0.0002152416741149
11	0.0001280502823882
12	-0.0000201348547807
13	-0.0000012504934821
14	0.0000011330272320
15	-0.0000002056338417
16	0.0000000061160950
17	0.0000000050020075
18	-0.0000000011812746
19	0.0000000001043427
20	0.0000000000077823
21	-0.0000000000036968
22	0.0000000000005100
23	-0.0000000000000206
24	-0.0000000000000054
25	0.0000000000000014
26	0.0000000000000001

APPENDIX B

Summary Statistical Tabulation
for the Accuracy of Methods

Code	ES3	EP3	DV3	LA3	HT3	ES2	EP2	DV2	LA2	DV1	DVIA
Method	Exponentiated Summation	Euler Product	Davis Series	Lanczos	Hastings Type II	Exponentiated Summation	Euler Product	Davis Series	Lanczos	Davis Series	Davis Series
Class	3	3	3	3	3	2	2	2	2	1	1
Range (z)	0.0001-1	0.0001-1	0.0001-1	0.0001-1	0.0001-1	1-2	1-2	1-2	1-2	1-12	1-12
n	101	101	101	101	101	101	101	101	101	101	101
$\Sigma\delta^2$	1.22190044E-02	6.21295207E-02	7.25145625E-13	4.66820088E+04	9.23606875E-03	2.00635709E+57	9.78822835E-01	1.44286189E-15	3.19945270E-01	3.85519896E+15	3.85519896E+15
RMS	1.09991018E-02	2.48020920E-02	8.47328723E-08	2.14987932E+01	9.56275197E-03	4.45700815E+27	9.84444777E-02	3.77965095E-09	5.62829899E-02	6.17821064E+06	6.17821064E+06
s	3.82612909E-04	5.84230904E-04	7.21434114E-14	4.64423881E+03	2.39368147E-04	1.99387555E+56	6.93952063E-03	5.21403829E-17	8.39110156E-04	1.92775322E+14	1.92775322E+14
$\sigma(n)^{0.5}$	3.80714074E-04	5.81331477E-04	7.17853774E-14	4.62119034E+03	2.38186208E-04	1.98398032E+56	6.90508110E-03	5.18816201E-17	8.34945811E-04	1.91818615E+14	1.90866656E+13
MEF	2.36757615E+00	1.81363492E+00	1.01962226E+00	1.45719177E+00	2.36929633E+00	-2.66624231E+00	1.00106371E+00	8.85726984E+00	1.21877323E+00	1.27234809E+00	3.30696923E-02
$\text{MAX}(d_L)$	4.84294024E+00	-5.60984333E-08	4.17155199E-09	-2.16054366E+00	2.16522709E-05	4.47643127E+30	-4.50942970E-01	5.89255461E-09	-4.81346791E+00	5.87840806E-09	-4.38614867E+00
$\text{MIN}(d_L)$	-4.07335819E-03	-4.50942970E+00	-8.53701638E-09	-4.81346791E+00	-3.58683430E+00	4.84294024E+00	-1.64142174E+01	-1.87985046E+02	-7.20153009E+00	-1.00000000E+02	-1.00000000E+02

Log_{128}(statistics)

	ES3	EP3	DV3	LA3	HT3	ES2	EP2	DV2	LA2	DV1	DVIA
n	9.51173069E-01	9.51173069E-01	9.51173069E-01	9.51173069E-01	9.51173069E-01	9.51173069E-01	9.51173069E-01	9.51173069E-01	9.51173069E-01	9.51173069E-01	9.51173069E-01
$\Sigma\delta^2$	-9.07818493E-01	-5.72653895E-01	-5.76097064E+00	2.21579200E+00	-9.65560768E-01	2.71934971E+01	-4.41476606E-03	-7.04285546E+00	-2.34871851E-01	7.39653239E+00	7.39653239E+00
RMS	-9.29495781E-01	-7.61913482E-01	-3.35607185E+00	6.32311967E-01	-9.58336918E-01	1.31211620E+01	-4.77792273E-01	-3.99701426E+00	-5.93022460E-01	3.22267966E+00	3.22267966E+00
s	-1.62168955E+00	-1.53445339E+00	-6.23658934E+00	1.74017521E+00	-1.71835446E+00	2.67176490E+01	-1.02442118E+00	-7.72719663E+00	-1.45983602E+00	6.77810515E+00	6.77913053E+00
$\sigma(n)^{0.5}$	-1.62271493E+00	-1.53547876E+00	-6.23761476E+00	1.73914983E+00	-1.71937983E+00	2.67166244E+01	-1.02546656E+00	-7.72822201E+00	-1.46086140E+00	6.77810515E+00	6.77810515E+00
MEF	-2.09830146E+00	-2.01106530E+00	-6.71320130E+00	1.26356329E+00	-2.19496637E+00	2.62410379E+01	-1.50103310E+00	-8.20380855E+00	-1.93644794E+00	6.30251862E+00	6.30251862E+00
$\text{MAX}(d_L)$	3.25126172E-01	3.44106633E+00	-3.97668123E+00	-1.58770626E-01	-2.21235888E+00	1.45457417E+01	-3.10420711E-01	-3.90549423E+00	-3.23868096E-01	-3.90598962E+00	-3.04707816E-01
$\text{MIN}(d_L)$	1.13422366E+00	-3.10420713E-01	3.82908870E+00	-3.23868096E-01	-2.63244442E-01	3.25126172E-01	-5.76696294E-01	2.71727867E+00	-4.06900495E-01	-9.49122313E-01	-9.49122313E-01

Method	Davis Series	Lanczos	Stirling Formula	Extended Stirling Formula	Warren Type I	Warren Type II	Burnside	Windshitl
Code	DV1A	LA1	S11	SE1	WO1	W11	BS1	WS1
Class	1	1	1	1	1	1	1	1
Range (z)	1-12	1-12	1-12	1-12	1-12	1-12	1-12	1-12
n	101	101	101	101	101	101	101	101
$\Sigma\delta^2$	3.85519896E+15	7.13279148E+13	1.89645776E+11	2.08140049E-04	3.79933946E+07	3.04430387E+08	4.40887418E+10	1.86559844E+11
RMS	6.17821064E+06	8.40367169E+05	4.33322161E+04	1.43554609E-03	6.13328806E+02	1.73613428E+03	2.08931136E+04	4.29782173E+04
s	1.92775322E+14	3.56366715E+12	9.31373734E+09	1.03128527E-05	1.88585649E+06	1.49355136E+07	2.16669462E+09	9.16503248E+09
σ	1.91818615E+14	3.54598135E+12	9.26751504E+09	1.02616720E-05	1.87649735E+06	1.48613914E+07	2.15594173E+09	9.11954816E+09
$\sigma/(n)^{0.5}$	1.90866656E+13	3.52838332E+11	9.22152212E+08	1.02107453E-06	1.86718465E+05	1.47876372E+06	2.14524220E+08	9.07428958E+08
MEF	3.30696923E-02	9.36775087E-01	1.72868896E+00	5.18622799E+00	3.76607475E+00	3.01234534E+00	2.09117971E+00	1.85529088E+00
MAX(d_i)	-4.38614867E+00	-4.81346791E+00	-6.91879406E-01	2.86759054E-02	1.07368910E-02	7.53055839E-01	2.75077351E+00	1.76150581E+00
MIN(d_i)	-1.00000000E+02	-1.36384082E+01	-7.78629911E+00	2.30849325E-08	-3.87286397E+00	2.76988620E-02	3.33764920E-01	-2.88556869E+00

Log_128(statistics)

	Davis Series	Lanczos	Stirling Formula	Extended Stirling Formula	Warren Type I	Warren Type II	Burnside	Windshitl
n	9.51173069E-01	9.51173069E-01	9.51173069E-01	9.51173069E-01	9.51173069E-01	9.51173069E-01	9.51173069E-01	9.51173069E-01
$\Sigma\delta^2$	7.39653239E+00	6.57421886E+00	5.35207375E+00	-1.74716540E+00	3.59703504E+00	4.02593388E+00	5.05138446E+00	5.34869250E+00
RMS	3.22267966E+00	2.81152290E+00	2.20045034E+00	-1.34916923E+00	1.32293099E+00	1.53738040E+00	2.05010570E+00	2.19875972E+00
s	6.77913053E+00	5.95664282E+00	4.73095901E+00	-2.36645672E+00	2.97811264E+00	3.40406622E+00	4.43040695E+00	4.72764184E+00
σ	6.77810515E+00	5.95561744E+00	4.72993363E+00	-2.36748209E+00	2.97708726E+00	3.40358084E+00	4.42938158E+00	4.72661646E+00
$\sigma/(n)^{0.5}$	6.30251862E+00	5.48003091E+00	4.25434709E+00	-2.84306863E+00	2.50150073E+00	2.92799430E+00	3.95379504E+00	4.25102992E+00
MAX(d_i)	-3.04707816E-01	-3.23868096E-01	7.59153565E-02	-7.32002450E-01	-9.34685556E-01	-5.84530358E-02	2.08548194E-01	1.16687033E-01
MIN(d_i)	-9.49122313E-01	-5.38514766E-01	-4.22991109E-01	-3.62406761E+00	1.95503895E-01	-7.39147069E-01	-2.26156538E-01	-2.18407954E-01

APPENDIX C

The Methods Timing Program Project
SPEED.vbp

GAMMA
Gamma Solution

Page 117 of 214

13:50 Friday, 05 August 2022

```vbnet
Option Explicit

Private Declare Sub keybd_event Lib "user32" (ByVal bVk As Byte, ByVal _
    bScan As Byte, ByVal dwFlags As Long, ByVal dwExtraInfo As Long)

Private Declare Function GetVersionExA Lib "kernel32" _
    (lpVersionInformation As OSVERSIONINFO) As Integer

Private Type OSVERSIONINFO
    dwOSVersionInfoSize As Long
    dwMajorVersion As Long
    dwMinorVersion As Long
    dwBuildNumber As Long
    dwPlatformId As Long
    szCSDVersion As String * 128
End Type

Private Const KEYEVENTF_KEYUP = &H2
Private Const VK_SNAPSHOT = &H2C
Private Const VK_MENU = &H12

Dim blnAboveVer4 As Boolean

    Private Function CaptureScreen()
    If blnAboveVer4 Then
        keybd_event VK_SNAPSHOT, 0, 0, 0
     Else
        keybd_event VK_SNAPSHOT, 1, 0, 0
    End If
    End Function

    Private Function CaptureForm()
    If blnAboveVer4 Then
        keybd_event VK_SNAPSHOT, 1, 0, 0
     Else
        keybd_event VK_MENU, 0, 0, 0
        keybd_event VK_SNAPSHOT, 0, 0, 0
        keybd_event VK_SNAPSHOT, 0, KEYEVENTF_KEYUP, 0
        keybd_event VK_MENU, 0, KEYEVENTF_KEYUP, 0
    End If
    End Function

    Private Sub SaveBitMap(SpedInputFormFileName)
' Load the captured image into a PictureBox and print it
    InputPicture.Picture = Clipboard.GetData()
    SavePicture InputPicture.Picture, SpedInputFormFileName + ".bmp"
' Clipboard.Clear
'   Picture1.Picture = Clipboard.GetData()
'   Printer.PaintPicture Picture1.Picture, 0, 0
'   Printer.EndDoc
    End Sub


    Private Sub Exit_Click()
    Unload SpedOutputForm
    Unload SpedServiceForm
    Unload SpedPlotForm
    Unload Me
    End Sub

    Private Sub Form_Load()
    Dim osinfo As OSVERSIONINFO
    Dim retvalue As Integer
    osinfo.dwOSVersionInfoSize = 148
```

```
osinfo.szCSDVersion = Space$(128)
retvalue = GetVersionExA(osinfo)
If osinfo.dwMajorVersion > 4 Then blnAboveVer4 = True
InputPicture.Visible = True
End Sub

Private Sub Compute_Click()
Call DoCompute
End Sub
```

```
Option Explicit

Private Declare Sub keybd_event Lib "user32" (ByVal bVk As Byte, ByVal _
    bScan As Byte, ByVal dwFlags As Long, ByVal dwExtraInfo As Long)

Private Declare Function GetVersionExA Lib "kernel32" _
    (lpVersionInformation As OSVERSIONINFO) As Integer

Private Type OSVERSIONINFO
    dwOSVersionInfoSize As Long
    dwMajorVersion As Long
    dwMinorVersion As Long
    dwBuildNumber As Long
    dwPlatformId As Long
    szCSDVersion As String * 128
End Type

Private Const KEYEVENTF_KEYUP = &H2
Private Const VK_SNAPSHOT = &H2C
Private Const VK_MENU = &H12

Dim blnAboveVer4 As Boolean

    Private Function CaptureScreen()
    If blnAboveVer4 Then
        keybd_event VK_SNAPSHOT, 0, 0, 0
    Else
        keybd_event VK_SNAPSHOT, 1, 0, 0
    End If
    End Function

    Private Function CaptureForm()
    If blnAboveVer4 Then
        keybd_event VK_SNAPSHOT, 1, 0, 0
    Else
        keybd_event VK_MENU, 0, 0, 0
        keybd_event VK_SNAPSHOT, 0, 0, 0
        keybd_event VK_SNAPSHOT, 0, KEYEVENTF_KEYUP, 0
        keybd_event VK_MENU, 0, KEYEVENTF_KEYUP, 0
    End If
    End Function

    Private Sub SaveBitMap(SpedOutputFormFileName)
    Dim OutputPicture As Image
    OutputPicture.Picture = Clipboard.GetData()
    SavePicture OutputPicture.Picture, SpedOutputFormFileName + ".bmp"
'    SavePicture Clipboard.GetData(), SpedOutputFormFileName + ".bmp"
    End Sub

    Private Sub Form_Load()
    Dim osinfo As OSVERSIONINFO
    Dim retvalue As Integer
    osinfo.dwOSVersionInfoSize = 148
    osinfo.szCSDVersion = Space$(128)
    retvalue = GetVersionExA(osinfo)
    If osinfo.dwMajorVersion > 4 Then blnAboveVer4 = True
'    OutputPicture.Visible = True
    End Sub
```

```
Option Explicit

Private Declare Sub keybd_event Lib "user32" (ByVal bVk As Byte, ByVal _
    bScan As Byte, ByVal dwFlags As Long, ByVal dwExtraInfo As Long)

Private Declare Function GetVersionExA Lib "kernel32" _
    (lpVersionInformation As OSVERSIONINFO) As Integer

Private Type OSVERSIONINFO
    dwOSVersionInfoSize As Long
    dwMajorVersion As Long
    dwMinorVersion As Long
    dwBuildNumber As Long
    dwPlatformId As Long
    szCSDVersion As String * 128
End Type

Private Const KEYEVENTF_KEYUP = &H2
Private Const VK_SNAPSHOT = &H2C
Private Const VK_MENU = &H12

Dim blnAboveVer4 As Boolean

Private Function CaptureScreen()
    If blnAboveVer4 Then
        keybd_event VK_SNAPSHOT, 0, 0, 0
    Else
        keybd_event VK_SNAPSHOT, 1, 0, 0
    End If
End Function

Private Function CaptureForm()
    If blnAboveVer4 Then
        keybd_event VK_SNAPSHOT, 1, 0, 0
    Else
        keybd_event VK_MENU, 0, 0, 0
        keybd_event VK_SNAPSHOT, 0, 0, 0
        keybd_event VK_SNAPSHOT, 0, KEYEVENTF_KEYUP, 0
        keybd_event VK_MENU, 0, KEYEVENTF_KEYUP, 0
    End If
End Function

Private Sub SaveBitMap(SpedInputFormFileName)
' Load the captured image into a PictureBox and print it
    Picture1.Picture = Clipboard.GetData()
    SavePicture Picture1.Picture, SpedInputFormFileName + ".bmp"
'   Picture1.Picture = Clipboard.GetData()
'   Printer.PaintPicture Picture1.Picture, 0, 0
'   Printer.EndDoc
End Sub

    Private Sub Form_Load()
    Dim osinfo As OSVERSIONINFO
    Dim retvalue As Integer
    osinfo.dwOSVersionInfoSize = 148
    osinfo.szCSDVersion = Space$(128)
    retvalue = GetVersionExA(osinfo)
    If osinfo.dwMajorVersion > 4 Then blnAboveVer4 = True
    Picture1.Visible = True
    End Sub

    Private Sub Quit_Click()
' A Subroutine to Unload the Project
    SpedInputForm.Hide
```

```vb
        SpedOutputForm.Hide
        SpedServiceForm.Hide
        SpedPlotForm.Hide
        Unload Me
        End Sub

        Private Sub Size_DblClick()
' A Subroutine to Adjust the Forms' Size
        Dim MP As Integer
        MP = SpedServiceForm.Size
        Cantonise (MP)
        End Sub

        Private Sub SpedInputFormPrint_Click()
' A Subroutine to Print the Input Form
        SpedInputForm.PrintForm
        End Sub

        Private Sub SpedOutputFormPrint_Click()
' A Subroutine to Print the Output Form
        SpedOutputForm.PrintForm
        End Sub

        Private Sub SpedPlotFormPrint_Click()
' A Subroutine to Print the Plot Form
        SpedPlotForm.PrintForm
        End Sub

        Private Sub SpedInputFormSave_Click()
' A Subroutine to Save a Bitmap Image of the Input Form
        DoEvents
        SpedInputForm.SetFocus
        Call CaptureForm
        DoEvents
        Call CaptureForm
        DoEvents
        Call SaveBitMap(SpedInputFormFileName)
        SpedServiceForm.SetFocus
        End Sub

        Private Sub SpedOutputFormSave_Click()
' A Subroutine to Save the Output Form
        DoEvents
        SpedOutputForm.SetFocus
        Call CaptureForm
        DoEvents
        Call CaptureForm
        DoEvents
        Picture1.Picture = Clipboard.GetData()
        SavePicture Picture1.Picture, SpedOutputFormFileName + ".bmp"
        SpedServiceForm.SetFocus
        End Sub

        Private Sub SpedPlotFormSave_Click()
' A Subroutine to Save the Plot Form
        SavePicture SpedPlotForm.Image, SpedPlotFormFileName + ".bmp"
        End Sub
```

```
' VARIABLE TYPE DEFAULTS
    DefDbl A-H, O-R, T-Z
    DefStr S
    DefInt I-K, M-N
    DefLng L

    Public Sub DoCompute()
' Assign Service Form Product Bitmap Filenames
    SpedServiceForm.SpedInputFormFileName = SpedInputForm.FileName + "In"
    SpedServiceForm.SpedOutputFormFileName = SpedInputForm.FileName + "Out"
    SpedServiceForm.SpedPlotFormFileName = SpedInputForm.FileName + "Plot"
' Assign the Sequential Files' Transput Parameters
'   File One ( The Job's Descriptors Input File )
    ILogicUnit(1) = 1
    STransmode(1) = "Input"
    SPath(1) = SpedInputForm.Path
    SFile(1) = SpedInputForm.FileName
    SExt(1) = ".JBS"
'   File Two ( The Methods' Constants Input File )
    ILogicUnit(2) = 2
    STransmode(2) = "Input"
    SPath(2) = SpedInputForm.Path
    SFile(2) = SpedInputForm.FileName
    SExt(2) = ".KON"
'   File Three ( The Ranges Input File )
    ILogicUnit(3) = 3
    STransmode(3) = "Input"
    SPath(3) = SpedInputForm.Path
    SFile(3) = SpedInputForm.FileName
    SExt(3) = ".RNG"
'   File Four ( The Method Results Output Files )
    ILogicUnit(4) = 4
    STransmode(4) = "Output"
    SPath(4) = SpedInputForm.Path
    SFile(4) = SpedInputForm.FileName
    SExt(4) = ".MRF"
' Load Constants
    NC = SpedInputForm.NC
'     Call CONSTIN(NM, MN(), SMC(), SMD(), DLB(), DUB(), NV(), NMC(), C())
    Call JOBSIN(NJ, IJ(), ICS(), IRS(), MN(), SMC(), SMD())
    Call KONSTIN(NK, IK(), NMC(), C())
    Call RANGIN(NR, IR(), DLB(), DUB(), NV())
' Do Timings
    DStart = Timer
' ** Job 1 Block ( Davis Sum by Power Series )
    SpedOutputForm.Status = SMD(1) + " BEGUN"
    Call METHODHOLDERxDP(NC, 1, 1, 3, SMC(), DLB(), DUB(), NV(), NMC(), Z(), G(), C(), T())
    SpedOutputForm.Status = SMD(1) + " FINISHED"
' ** Job 2 Block ( Davis Sum by Product Series )
    SpedOutputForm.Status = SMD(2) + " BEGUN"
    Call METHODHOLDERxDR(NC, 2, 1, 3, SMC(), DLB(), DUB(), NV(), NMC(), Z(), G(), C(), T())
    SpedOutputForm.Status = SMD(2) + " FINISHED"
' ** Job 3 Block ( Exponentiated Summation )
    SpedOutputForm.Status = SMD(3) + " BEGUN"
    Call METHODHOLDERxES(NC, 3, 0, 3, SMC(), DLB(), DUB(), NV(), NMC(), Z(), G(), C(), T())
    SpedOutputForm.Status = SMD(3) + " FINISHED"
' ** Job 4 Block ( Euler Product )
    SpedOutputForm.Status = SMD(4) + " BEGUN"
    Call METHODHOLDERxEP(NC, 4, 0, 3, SMC(), DLB(), DUB(), NV(), NMC(), Z(), G(), C(), T())
    SpedOutputForm.Status = SMD(4) + " FINISHED"
' ** Job 5 Block ( Lanczos Approximation )
    SpedOutputForm.Status = SMD(5) + " BEGUN"
    Call METHODHOLDERxLA(NC, 5, 5, 3, SMC(), DLB(), DUB(), NV(), NMC(), Z(), G(), C(), T())
    SpedOutputForm.Status = SMD(5) + " FINISHED"
```

```
' ** Job 6 Block ( Hastings Type II )
    SpedOutputForm.Status = SMD(6) + " BEGUN"
    Call METHODHOLDERxHT(NC, 6, 7, 3, SMC(), DLB(), DUB(), NV(), NMC(), Z(), G(), C(), T())
    SpedOutputForm.Status = SMD(6) + " FINISHED"
' ** Job 7 Block ( Davis Sum by Power Series )
    SpedOutputForm.Status = SMD(7) + " BEGUN"
    Call METHODHOLDERxDP(NC, 7, 1, 2, SMC(), DLB(), DUB(), NV(), NMC(), Z(), G(), C(), T())
    SpedOutputForm.Status = SMD(7) + " FINISHED"
' ** Job 8 Block ( Davis Sum by Product Series )
    SpedOutputForm.Status = SMD(8) + " BEGUN"
    Call METHODHOLDERxDR(NC, 8, 1, 2, SMC(), DLB(), DUB(), NV(), NMC(), Z(), G(), C(), T())
    SpedOutputForm.Status = SMD(8) + " FINISHED"
' ** Job 9 Block ( Exponentiated Summation )
    SpedOutputForm.Status = SMD(9) + " BEGUN"
    Call METHODHOLDERxES(NC, 9, 0, 2, SMC(), DLB(), DUB(), NV(), NMC(), Z(), G(), C(), T())
    SpedOutputForm.Status = SMD(9) + " FINISHED"
' ** Job 10 Block ( Euler Product )
    SpedOutputForm.Status = SMD(10) + " BEGUN"
    Call METHODHOLDERxEP(NC, 10, 0, 2, SMC(), DLB(), DUB(), NV(), NMC(), Z(), G(), C(), T())
    SpedOutputForm.Status = SMD(10) + " FINISHED"
' ** Job 11 Block ( Lanczos Approximation )
    SpedOutputForm.Status = SMD(11) + " BEGUN"
    Call METHODHOLDERxLA(NC, 11, 5, 2, SMC(), DLB(), DUB(), NV(), NMC(), Z(), G(), C(), T())
    SpedOutputForm.Status = SMD(11) + " FINISHED"
' ** Job 12 Block ( Davis Sum by Power Series )
    SpedOutputForm.Status = SMD(12) + " BEGUN"
    Call METHODHOLDERxDP(NC, 12, 1, 1, SMC(), DLB(), DUB(), NV(), NMC(), Z(), G(), C(), T())
    SpedOutputForm.Status = SMD(12) + " FINISHED"
' ** Job 13 Block ( Davis Sum by Product Series )
    SpedOutputForm.Status = SMD(13) + " BEGUN"
    Call METHODHOLDERxDR(NC, 13, 1, 1, SMC(), DLB(), DUB(), NV(), NMC(), Z(), G(), C(), T())
    SpedOutputForm.Status = SMD(13) + " FINISHED"
' ** Job 14 Block ( Lanczos Approximation )
    SpedOutputForm.Status = SMD(14) + " BEGUN"
    Call METHODHOLDERxLA(NC, 14, 5, 1, SMC(), DLB(), DUB(), NV(), NMC(), Z(), G(), C(), T())
    SpedOutputForm.Status = SMD(14) + " FINISHED"
' ** Job 15 Block ( Stirling Formula )
    SpedOutputForm.Status = SMD(15) + " BEGUN"
    Call METHODHOLDERxST(NC, 15, 0, 1, SMC(), DLB(), DUB(), NV(), NMC(), Z(), G(), C(), T())
    SpedOutputForm.Status = SMD(15) + " FINISHED"
' ** Job 16 Block ( Extended Stirling Formula )
    SpedOutputForm.Status = SMD(16) + " BEGUN"
    Call METHODHOLDERxSE(NC, 16, 2, 1, SMC(), DLB(), DUB(), NV(), NMC(), Z(), G(), C(), T())
    SpedOutputForm.Status = SMD(16) + " FINISHED"
' ** Job 17 Block ( Warren Type I )
    SpedOutputForm.Status = SMD(17) + " BEGUN"
    Call METHODHOLDERxWO(NC, 17, 0, 1, SMC(), DLB(), DUB(), NV(), NMC(), Z(), G(), C(), T())
    SpedOutputForm.Status = SMD(17) + " FINISHED"
' ** Job 18 Block ( Warren Type II [ Range One Configuration ] )
    SpedOutputForm.Status = SMD(18) + " BEGUN"
    Call METHODHOLDERxWT(NC, 18, 4, 1, SMC(), DLB(), DUB(), NV(), NMC(), Z(), G(), C(), T())
    SpedOutputForm.Status = SMD(18) + " FINISHED"
' ** Job 19 Block ( Burnside Formula )
    SpedOutputForm.Status = SMD(19) + " BEGUN"
    Call METHODHOLDERxBS(NC, 19, 0, 1, SMC(), DLB(), DUB(), NV(), NMC(), Z(), G(), C(), T())
    SpedOutputForm.Status = SMD(19) + " FINISHED"
' ** Job 20 Block ( Windshitl Formula )
    SpedOutputForm.Status = SMD(20) + " BEGUN"
    Call METHODHOLDERxWS(NC, 20, 0, 1, SMC(), DLB(), DUB(), NV(), NMC(), Z(), G(), C(), T())
    SpedOutputForm.Status = SMD(20) + " FINISHED"
' ** End of Job Blocks
    DFinish = Timer
    SpedOutputForm.PMT = DFinish - DStart
' Terminate
```

```
SpedOutputForm.Status = "ALL FINISHED"
Close
End Sub
```

```
' VARIABLE TYPE DEFAULTS
    DefDbl A-H, O-R, T-Z
    DefStr S
    DefInt I-K, M-N
    DefLng L
' Object Definitions
'    ( None )
' Publicise Object Instantiations
'    ( None )
' Static Array Declarations
'    ( None )
' Pseudo-Static Array Declarations
    Public ILogicUnit(), STransmode(), SPath(), SFile(), SExt()
' Dynamic Array Declarations
    Public IJ(), IK(), IR()
    Public ICS(), IRS()
    Public MN(), SMC(), SMD(), DLB(), DUB(), NV(), NMC()
    Public C()
    Public Z(), G(), T()
' Constant Definitions
    Const PI As Double = 3.14159265358979
    Const PI2 As Double = 6.28318530717959
    Const HP As Double = 1.5707963267949
    Const RC As Double = 2.506628274631
    Const EMC As Double = 0.577215664901533
    Const ENAP As Double = 2.71828182845905
' Declare Logical Unit Number Holders
    Public IU, IV, IW, IX
' Declare String Constant Data ( for Common Shared emulation )
    Public SC, SM, SCR
' Declare Formats ( for Common Shared emulation )
    Public SI4, SF10P8, SF11P6
' Declare Dynamic Array Dimensions
    Public NJX, NKX, NRX, NMX, NVX, NMCX, NCX
' Declare Plotting Parameters
'    ( None )

    Sub Main()
'    Program SPEED
' A Program to Cantonise Input, Output, Service and Plot Forms; and
' To Compute a Set of Empirical Operation and Function Mill Times
' Based upon Data Located from SpedInputForm
'    Written by:-
'
'        James R Warren BSc MSc PhD PGCE
'        "Southgate"
'        31 Victoria Avenue
'        Bloxwich
'        Walsall
'        WS3 3HS
'
'        5 September 2004
'
'    This Program is Written in MicroSoft Visual Basic 6.0
'
' Assign the Pseudo-Static Array Dimensions Maxima and Dimension the Pseudo-Static Arrays
    IModes = 3: ITimes = 2
    ReDim T(IModes, ITimes)
    NUnits = 10
    ReDim ILogicUnit(NUnits), STransmode(NUnits), SPath(NUnits), SFile(NUnits), SExt(NUnits)
' Assign the Dynamic Arrays' Maxima and Dimension the Dynamic Arrays
    NJX = 100: NKX = 100: NRX = 5
    NMX = 100: NVX = 1000: NMCX = 100: NCX = 5000
    ReDim IJ(NJX), IK(NKX), IR(NRX)
```

```
        ReDim ICS(NJX), IRS(NJX)
        ReDim MN(NMX), SMC(NMX), SMD(NMX), DLB(NMX), DUB(NMX), NV(NMX), NMC(NMX)
        ReDim C(NMX, NMCX)
        ReDim Z(NVX), G(NVX)
' Form Definitions
        Dim SpedInputForm As Form
        Dim SpedOutputForm As Form
        Dim SpedServiceForm As Form
        Dim SpedPlotForm As Form
' String Constant Definitions
        SC = ":": SM = ",": SCR = Chr(13) + Chr(10)
' Format Definitions
        SI4 = "####": SF10P8 = "#.00000000": SF11P6 = "####.000000"
' Set MarginPercent
        MP = 12
' Perform Actions
        Call Cantonise(MP)
' Terminate
        End Sub


        Public Sub Cantonise(MP)
' A Subroutine to Set Up Window's Cantonisation and Display; and to
' Correct Excessively Small Size Choices
'        Argument:-
'            MP     The Margin Percent Windows' Scale Factor
'
        If MP > 40 Then MP = 40
' Define Variables
        Dim InsetOne, InsetTwo, DownsetOne, DownsetTwo As Integer
        Const SQRT2 As Double = 1.4142135623731
' Define Form Dimensions
        SpedInputForm.Height = Screen.Height * (50 - MP) / 100
        SpedInputForm.Width = SQRT2 * SpedInputForm.Height
        SpedOutputForm.Height = SpedInputForm.Height
        SpedServiceForm.Height = SpedInputForm.Height
        SpedPlotForm.Height = SpedInputForm.Height
        SpedOutputForm.Width = SpedInputForm.Width
        SpedServiceForm.Width = SpedInputForm.Width
        SpedPlotForm.Width = SpedInputForm.Width
' Define Form Positions
        InsetOne = (Screen.Width - 2 * SpedInputForm.Width) / 2
        InsetTwo = InsetOne + SpedInputForm.Width
        DownsetOne = MP * Screen.Height / 100
        DownsetTwo = DownsetOne + SpedInputForm.Height
        SpedInputForm.Left = InsetOne
        SpedInputForm.Top = DownsetOne
        SpedOutputForm.Left = InsetTwo
        SpedOutputForm.Top = DownsetOne
        SpedServiceForm.Left = InsetOne
        SpedServiceForm.Top = DownsetTwo
        SpedPlotForm.Left = InsetTwo
        SpedPlotForm.Top = DownsetTwo
' Show the Forms
        SpedInputForm.Show
        SpedOutputForm.Show
        SpedServiceForm.Show
        SpedPlotForm.Show
        End Sub

        Public Sub FileOpener(IFS, ILU)
' A Subroutine to Open a Sequential Access File whose Transput Properties
' Are Stored in Public ( Shared ) Arrays
'        Arguments:-
'            IFS     The File Execution Serial Number
```

```
'       ILU   The File Logical Unit Number
' ( Arrays ILogicUnit, STransMode, SPath, SFile and SExt
'   are Common Shared by static Public Declaration )
'
      ILU = ILogicUnit(IFS)
      SP = SPath(IFS) + SFile(IFS) + SExt(IFS)
      STM = STransmode(IFS)
      Select Case STM
        Case "Input"
          Open SP For Input As #ILU
        Case "Output"
          Open SP For Output As #ILU
        Case "Append"
          Open SP For Append As #ILU
      End Select
      End Sub

      Public Sub PrintAnywhere(Src As Object, Dest As Object)
' A Subroutine to Mediate Graphics Presentation upon a Printed Form
      Dest.PaintPicture Src.Picture, Dest.Width / 2, Dest.Height / 2
      If Dest Is Printer Then
        Printer.EndDoc
      End If
      End Sub

      Public Sub CONSTIN(NM, MN(), SMC(), SMD(), DLB(), DUB(), NV(), NMC(), C())
' A A Subroutine to Input Descriptor, Range and Constants Data ( if applicable )
' For All the Timed Methods
'     Arguments:-
'       NM    The Number of Methods
'       MN()  The Array of Method Numbers
'       SMC() The Method Code
'       SMD() The Method Descriptor
'       DLB() The Range Lower Bound
'       DUB() The Range Upper Bound
'       NV()  The Number of Intervals
'       NMC() The Number of Method Constants
'       C()   The Method Constants Array
' ( Arrays ILogicUnit, STransmode, SPath, SFile and SExt
'   are Common Shared by static Public Declaration )
'
' Open the File
      Call FileOpener(1, IU)
' Input the Data
      Input #IU, NM
      For I = 1 To NM
        Input #IU, MN(I)
        Input #IU, SMC(I)
        Input #IU, SMD(I)
        Input #IU, DLB(I)
        Input #IU, DUB(I)
        Input #IU, NV(I)
        Input #IU, NMC(I)
        For J = 1 To NMC(I)
          Input #IU, C(J, MN(I))
        Next J
      Next I
' Terminate
      Close 1
      End Sub

      Public Sub JOBSIN(NJ, IJ(), ICS(), IRS(), MN(), SMC(), SMD())
' A Subroutine to Load the Job's Descriptors Input Data
'     Arguments:-
```

```
'          NJ    The Number of Jobs
'          IJ()  The Array of Job Serial Numbers
'          ICS() The Array of Constants Selector Numbers
'          IRS() The Array of Range    Selector Numbers
'          MN()  The Array of Method Numbers
'          SMC() The Array of Method Codes
'          SMD() The Array of Method Descriptions
' Open the File
    Call FileOpener(1, IU)
' Input the Data
    Input #IU, NJ
    For I = 1 To NJ
    Input #IU, IJ(I), ICS(I), IRS(I), MN(I), SMC(I), SMD(I)
    Next I
' Terminate
    Close IU
    End Sub

    Public Sub KONSTIN(NK, IK(), NMC(), C())
' A Subroutine to Load the Method's Constants Input Data
'     Arguments:-
'       NK    The Number of Sets of Constants
'       IK()  The Array of Constants' Set Serials
'       NMC() The Number of Method Constants
'       C()   The Array  of Method Constants
'
' Open the File
    Call FileOpener(2, IU)
' Input the Data
    Input #IU, NK
    For I = 1 To NK
    Input #IU, IK(I), NMC(I)
    For J = 1 To NMC(I)
      Input #IU, C(IK(I), J)
    Next J
    Next I
' Terminate
    Close IU
    End Sub

    Public Sub RANGIN(NR, IR(), DLB(), DUB(), NV())
' A Subroutine to Load the Ranges Input Data
'     Arguments:-
'       NR    The Number of Ranges
'       IR()  The Array of Range Serials
'       DLB() The Array of Range Lower Bounds
'       DUB() The Array of Range Upper Bounds
'       NV()  The Array of Numbers of Intervals
'
' Open the File
    Call FileOpener(3, IU)
' Input the Data
    Input #IU, NR
    For I = 1 To NR
    Input #IU, IR(I), DLB(I), DUB(I), NV(I)
    Next I
' Terminate
    Close IU
    End Sub

    Public Sub TIMEOUT(JN, JR, SMC(), NV(), Z(), G(), T())
' A Subroutine to Record an Operation's or Method's
' Algorithm and Carcass Elaboration Times and
' Individual Argument Increments with their Point Solutions
```

```
'       Arguments:-
'           JN     The Job Number
'           JR     The Range Number
'           SMC()  The Method Code
'           NV()   The Number of Intervals
'           Z()    The Array of Method Arguments
'           G()    The Array of Method Solutions
'           T()    The Array of Elaboration Times
' ( Arrays ILogicUnit, STransmode, SPath, SFile and SExt
'   are Common Shared by static Public Declaration )
'
' Open the File
        IFS = 4
        IV = ILogicUnit(IFS)
        SFile(IFS) = SMC(JN)
        SP = SPath(IFS) + SFile(IFS) + SExt(IFS)
        Open SP For Output As #IV
' Write the Header
        Write #IV,
        Write #IV, "Interval Number", "z", SMC(JN) + " G(z)"
' Write Point Solutions
        For I = 0 To NV(JR)
          Write #IV, I, Z(I), G(I)
        Next I
        Write #IV,
' Write Elaboration Times
        Write #IV, "Composite Start Time", "", T(1, 1)
        Write #IV, "Composite Finish Time", "", T(1, 2)
        Write #IV,
        Write #IV, "Carcass Start Time", "", T(2, 1)
        Write #IV, "Carcass Finish Time", "", T(2, 2)
        Write #IV,
        Write #IV, "Method Mill Time", "", T(3, 1)
' Terminate
        Close IV
        End Sub


        Public Sub METHODHOLDERxDP(NC, JN, KN, JR, SMC(), DLB(), DUB(), NV(), NMC(), Z(), G(), C(),
T())
' A Subroutine to Invoke and Time an Operation or Method Elaboration
'       Arguments:-
'           NC     The Number of Cycles
'           JN     The Job Number
'           KN     The Constants' Set Number
'           JR     The Range Number
'           SMC()  The Method Code
'           DLB()  The Range Lower Bound
'           DUB()  The Range Upper Bound
'           NV()   The Number of Intervals
'           NMC()  The Number of Method Constants
'           Z()    The Array of Method Arguments
'           G()    The Array of Method Solutions
'           C()    The Array of Method Constants
'           T()    The Array of Times
'
' Set Local Constants
        NW = NV(JR)
        NX = NMC(KN)
        DD = DLB(JR)
        ZI = (DUB(JR) - DLB(JR)) / NV(JR)
' Do Composite
        T(1, 1) = Timer
        For I = 1 To NC
          For J = 0 To NW
```

```
        Z(J) = DD + J * ZI
        Call METHODxDP(KN, NX, Z(J), G(J), C())
      Next J
    Next I
    T(1, 2) = Timer
' Do Carcass
    T(2, 1) = Timer
    For I = 1 To NC
      For J = 0 To NW
        Z(J) = DD + J * ZI
        Call METHODxDPcarc(KN, NX, Z(J), G(J), C())
      Next J
    Next I
    T(2, 2) = Timer
' Calculate Mill Time
    T(3, 1) = (T(1, 2) - T(1, 1) - T(2, 2) + T(2, 1)) / NC
' Output Results
    Call TIMEOUT(JN, JR, SMC(), NV(), Z(), G(), T())
' Terminate
    End Sub

    Public Sub METHODHOLDERxDR(NC, JN, KN, JR, SMC(), DLB(), DUB(), NV(), NMC(), Z(), G(), C(),
T())
' A Subroutine to Invoke and Time an Operation or Method Elaboration
'    Arguments:-
'      NC    The Number of Cycles
'      JN    The Job Number
'      KN    The Constants' Set Number
'      JR    The Range Number
'      SMC() The Method Code
'      DLB() The Range Lower Bound
'      DUB() The Range Upper Bound
'      NV()  The Number of Intervals
'      NMC() The Number of Method Constants
'      Z()   The Array of Method Arguments
'      G()   The Array of Method Solutions
'      C()   The Array of Method Constants
'      T()   The Array of Times
'
' Set Local Constants
    NW = NV(JR)
    NX = NMC(KN)
    DD = DLB(JR)
    ZI = (DUB(JR) - DLB(JR)) / NV(JR)
' Do Composite
    T(1, 1) = Timer
    For I = 1 To NC
      For J = 0 To NW
        Z(J) = DD + J * ZI
        Call METHODxDR(KN, NX, Z(J), G(J), C())
      Next J
    Next I
    T(1, 2) = Timer
' Do Carcass
    T(2, 1) = Timer
    For I = 1 To NC
      For J = 0 To NW
        Z(J) = DD + J * ZI
        Call METHODxDRcarc(KN, NX, Z(J), G(J), C())
      Next J
    Next I
    T(2, 2) = Timer
' Calculate Mill Time
    T(3, 1) = (T(1, 2) - T(1, 1) - T(2, 2) + T(2, 1)) / NC
```

```
' Output Results
    Call TIMEOUT(JN, JR, SMC(), NV(), Z(), G(), T())
' Terminate
    End Sub

    Public Sub METHODHOLDERxES(NC, JN, KN, JR, SMC(), DLB(), DUB(), NV(), NMC(), Z(), G(), C(),
T())
' A Subroutine to Invoke and Time an Operation or Method Elaboration
'    Arguments:-
'    NC    The Number of Cycles
'    JN    The Job Number
'    KN    The Constants' Set Number
'    JR    The Range Number
'    SMC() The Method Code
'    DLB() The Range Lower Bound
'    DUB() The Range Upper Bound
'    NV()  The Number of Intervals
'    NMC() The Number of Method Constants
'    Z()   The Array of Method Arguments
'    G()   The Array of Method Solutions
'    C()   The Array of Method Constants
'    T()   The Array of Times
'
' Set Local Constants
    NW = NV(JR)
    NX = NMC(KN)
    DD = DLB(JR)
    ZI = (DUB(JR) - DLB(JR)) / NV(JR)
' Do Composite
    T(1, 1) = Timer
    For I = 1 To NC
      For J = 0 To NW
        Z(J) = DD + J * ZI
        Call METHODxES(KN, NX, Z(J), G(J), C())
      Next J
    Next I
    T(1, 2) = Timer
' Do Carcass
    T(2, 1) = Timer
    For I = 1 To NC
      For J = 0 To NW
        Z(J) = DD + J * ZI
        Call METHODxEScarc(KN, NX, Z(J), G(J), C())
      Next J
    Next I
    T(2, 2) = Timer
' Calculate Mill Time
    T(3, 1) = (T(1, 2) - T(1, 1) - T(2, 2) + T(2, 1)) / NC
' Output Results
    Call TIMEOUT(JN, JR, SMC(), NV(), Z(), G(), T())
' Terminate
    End Sub

    Public Sub METHODHOLDERxEP(NC, JN, KN, JR, SMC(), DLB(), DUB(), NV(), NMC(), Z(), G(), C(),
T())
' A Subroutine to Invoke and Time an Operation or Method Elaboration
'    Arguments:-
'    NC    The Number of Cycles
'    JN    The Job Number
'    KN    The Constants' Set Number
'    JR    The Range Number
'    SMC() The Method Code
'    DLB() The Range Lower Bound
'    DUB() The Range Upper Bound
```

```
'        NV()   The Number of Intervals
'        NMC()  The Number of Method Constants
'        Z()    The Array of Method Arguments
'        G()    The Array of Method Solutions
'        C()    The Array of Method Constants
'        T()    The Array of Times
'
' Set Local Constants
      NW = NV(JR)
      NX = NMC(KN)
      DD = DLB(JR)
      ZI = (DUB(JR) - DLB(JR)) / NV(JR)
' Do Composite
      T(1, 1) = Timer
      For I = 1 To NC
        For J = 0 To NW
          Z(J) = DD + J * ZI
          Call METHODxEP(KN, NX, Z(J), G(J), C())
        Next J
      Next I
      T(1, 2) = Timer
' Do Carcass
      T(2, 1) = Timer
      For I = 1 To NC
        For J = 0 To NW
          Z(J) = DD + J * ZI
          Call METHODxEPcarc(KN, NX, Z(J), G(J), C())
        Next J
      Next I
      T(2, 2) = Timer
' Calculate Mill Time
      T(3, 1) = (T(1, 2) - T(1, 1) - T(2, 2) + T(2, 1)) / NC
' Output Results
      Call TIMEOUT(JN, JR, SMC(), NV(), Z(), G(), T())
' Terminate
      End Sub

      Public Sub METHODHOLDERxLA(NC, JN, KN, JR, SMC(), DLB(), DUB(), NV(), NMC(), Z(), G(), C(),
T())
' A Subroutine to Invoke and Time an Operation or Method Elaboration
'      Arguments:-
'        NC     The Number of Cycles
'        JN     The Job Number
'        KN     The Constants' Set Number
'        JR     The Range Number
'        SMC()  The Method Code
'        DLB()  The Range Lower Bound
'        DUB()  The Range Upper Bound
'        NV()   The Number of Intervals
'        NMC()  The Number of Method Constants
'        Z()    The Array of Method Arguments
'        G()    The Array of Method Solutions
'        C()    The Array of Method Constants
'        T()    The Array of Times
'
' Set Local Constants
      NW = NV(JR)
      NX = NMC(KN)
      DD = DLB(JR)
      ZI = (DUB(JR) - DLB(JR)) / NV(JR)
' Do Composite
      T(1, 1) = Timer
      For I = 1 To NC
        For J = 0 To NW
```

```
          Z(J) = DD + J * ZI
          Call METHODxLA(KN, NX, Z(J), G(J), C())
        Next J
      Next I
      T(1, 2) = Timer
' Do Carcass
      T(2, 1) = Timer
      For I = 1 To NC
        For J = 0 To NW
          Z(J) = DD + J * ZI
          Call METHODxLAcarc(KN, NX, Z(J), G(J), C())
        Next J
      Next I
      T(2, 2) = Timer
' Calculate Mill Time
      T(3, 1) = (T(1, 2) - T(1, 1) - T(2, 2) + T(2, 1)) / NC
' Output Results
      Call TIMEOUT(JN, JR, SMC(), NV(), Z(), G(), T())
' Terminate
      End Sub


      Public Sub METHODHOLDERxHT(NC, JN, KN, JR, SMC(), DLB(), DUB(), NV(), NMC(), Z(), G(), C(),
T())
' A Subroutine to Invoke and Time an Operation or Method Elaboration
'     Arguments:-
'     NC    The Number of Cycles
'     JN    The Job Number
'     KN    The Constants' Set Number
'     JR    The Range Number
'     SMC() The Method Code
'     DLB() The Range Lower Bound
'     DUB() The Range Upper Bound
'     NV()  The Number of Intervals
'     NMC() The Number of Method Constants
'     Z()   The Array of Method Arguments
'     G()   The Array of Method Solutions
'     C()   The Array of Method Constants
'     T()   The Array of Times
'
' Set Local Constants
      NW = NV(JR)
      NX = NMC(KN)
      DD = DLB(JR)
      ZI = (DUB(JR) - DLB(JR)) / NV(JR)
' Do Composite
      T(1, 1) = Timer
      For I = 1 To NC
        For J = 0 To NW
          Z(J) = DD + J * ZI
          Call METHODxHT(KN, NX, Z(J), G(J), C())
        Next J
      Next I
      T(1, 2) = Timer
' Do Carcass
      T(2, 1) = Timer
      For I = 1 To NC
        For J = 0 To NW
          Z(J) = DD + J * ZI
          Call METHODxHTcarc(KN, NX, Z(J), G(J), C())
        Next J
      Next I
      T(2, 2) = Timer
' Calculate Mill Time
      T(3, 1) = (T(1, 2) - T(1, 1) - T(2, 2) + T(2, 1)) / NC
```

```
' Output Results
    Call TIMEOUT(JN, JR, SMC(), NV(), Z(), G(), T())
' Terminate
    End Sub

    Public Sub METHODHOLDERxST(NC, JN, KN, JR, SMC(), DLB(), DUB(), NV(), NMC(), Z(), G(), C(),
T())
' A Subroutine to Invoke and Time an Operation or Method Elaboration
'    Arguments:-
'    NC     The Number of Cycles
'    JN     The Job Number
'    KN     The Constants' Set Number
'    JR     The Range Number
'    SMC()  The Method Code
'    DLB()  The Range Lower Bound
'    DUB()  The Range Upper Bound
'    NV()   The Number of Intervals
'    NMC()  The Number of Method Constants
'    Z()    The Array of Method Arguments
'    G()    The Array of Method Solutions
'    C()    The Array of Method Constants
'    T()    The Array of Times
'
' Set Local Constants
    NW = NV(JR)
    NX = NMC(KN)
    DD = DLB(JR)
    ZI = (DUB(JR) - DLB(JR)) / NV(JR)
' Do Composite
    T(1, 1) = Timer
    For I = 1 To NC
      For J = 0 To NW
        Z(J) = DD + J * ZI
        Call METHODxST(KN, NX, Z(J), G(J), C())
      Next J
    Next I
    T(1, 2) = Timer
' Do Carcass
    T(2, 1) = Timer
    For I = 1 To NC
      For J = 0 To NW
        Z(J) = DD + J * ZI
        Call METHODxSTcarc(KN, NX, Z(J), G(J), C())
      Next J
    Next I
    T(2, 2) = Timer
' Calculate Mill Time
    T(3, 1) = (T(1, 2) - T(1, 1) - T(2, 2) + T(2, 1)) / NC
' Output Results
    Call TIMEOUT(JN, JR, SMC(), NV(), Z(), G(), T())
' Terminate
    End Sub

    Public Sub METHODHOLDERxSE(NC, JN, KN, JR, SMC(), DLB(), DUB(), NV(), NMC(), Z(), G(), C(),
T())
' A Subroutine to Invoke and Time an Operation or Method Elaboration
'    Arguments:-
'    NC     The Number of Cycles
'    JN     The Job Number
'    KN     The Constants' Set Number
'    JR     The Range Number
'    SMC()  The Method Code
'    DLB()  The Range Lower Bound
'    DUB()  The Range Upper Bound
```

```
'      NV()   The Number of Intervals
'      NMC()  The Number of Method Constants
'      Z()    The Array of Method Arguments
'      G()    The Array of Method Solutions
'      C()    The Array of Method Constants
'      T()    The Array of Times
'
' Set Local Constants
      NW = NV(JR)
      NX = NMC(KN)
      DD = DLB(JR)
      ZI = (DUB(JR) - DLB(JR)) / NV(JR)
' Do Composite
      T(1, 1) = Timer
      For I = 1 To NC
        For J = 0 To NW
          Z(J) = DD + J * ZI
          Call METHODxSE(KN, NX, Z(J), G(J), C())
        Next J
      Next I
      T(1, 2) = Timer
' Do Carcass
      T(2, 1) = Timer
      For I = 1 To NC
        For J = 0 To NW
          Z(J) = DD + J * ZI
          Call METHODxSEcarc(KN, NX, Z(J), G(J), C())
        Next J
      Next I
      T(2, 2) = Timer
' Calculate Mill Time
      T(3, 1) = (T(1, 2) - T(1, 1) - T(2, 2) + T(2, 1)) / NC
' Output Results
      Call TIMEOUT(JN, JR, SMC(), NV(), Z(), G(), T())
' Terminate
      End Sub


      Public Sub METHODHOLDERxWO(NC, JN, KN, JR, SMC(), DLB(), DUB(), NV(), NMC(), Z(), G(), C(),
T())
' A Subroutine to Invoke and Time an Operation or Method Elaboration
'    Arguments:-
'      NC     The Number of Cycles
'      JN     The Job Number
'      KN     The Constants' Set Number
'      JR     The Range Number
'      SMC()  The Method Code
'      DLB()  The Range Lower Bound
'      DUB()  The Range Upper Bound
'      NV()   The Number of Intervals
'      NMC()  The Number of Method Constants
'      Z()    The Array of Method Arguments
'      G()    The Array of Method Solutions
'      C()    The Array of Method Constants
'      T()    The Array of Times
'
' Set Local Constants
      NW = NV(JR)
      NX = NMC(KN)
      DD = DLB(JR)
      ZI = (DUB(JR) - DLB(JR)) / NV(JR)
' Do Composite
      T(1, 1) = Timer
      For I = 1 To NC
        For J = 0 To NW
```

```
            Z(J) = DD + J * ZI
            Call METHODxWO(KN, NX, Z(J), G(J), C())
          Next J
        Next I
        T(1, 2) = Timer
' Do Carcass
        T(2, 1) = Timer
        For I = 1 To NC
          For J = 0 To NW
            Z(J) = DD + J * ZI
            Call METHODxWOcarc(KN, NX, Z(J), G(J), C())
          Next J
        Next I
        T(2, 2) = Timer
' Calculate Mill Time
        T(3, 1) = (T(1, 2) - T(1, 1) - T(2, 2) + T(2, 1)) / NC
' Output Results
        Call TIMEOUT(JN, JR, SMC(), NV(), Z(), G(), T())
' Terminate
        End Sub


        Public Sub METHODHOLDERxWT(NC, JN, KN, JR, SMC(), DLB(), DUB(), NV(), NMC(), Z(), G(), C(),
T())
' A Subroutine to Invoke and Time an Operation or Method Elaboration
'       Arguments:-
'       NC      The Number of Cycles
'       JN      The Job Number
'       KN      The Constants' Set Number
'       JR      The Range Number
'       SMC()   The Method Code
'       DLB()   The Range Lower Bound
'       DUB()   The Range Upper Bound
'       NV()    The Number of Intervals
'       NMC()   The Number of Method Constants
'       Z()     The Array of Method Arguments
'       G()     The Array of Method Solutions
'       C()     The Array of Method Constants
'       T()     The Array of Times
'
' Set Local Constants
        NW = NV(JR)
        NX = NMC(KN)
        DD = DLB(JR)
        ZI = (DUB(JR) - DLB(JR)) / NV(JR)
' Do Composite
        T(1, 1) = Timer
        For I = 1 To NC
          For J = 0 To NW
            Z(J) = DD + J * ZI
            Call METHODxWT(KN, NX, Z(J), G(J), C())
          Next J
        Next I
        T(1, 2) = Timer
' Do Carcass
        T(2, 1) = Timer
        For I = 1 To NC
          For J = 0 To NW
            Z(J) = DD + J * ZI
            Call METHODxWTcarc(KN, NX, Z(J), G(J), C())
          Next J
        Next I
        T(2, 2) = Timer
' Calculate Mill Time
        T(3, 1) = (T(1, 2) - T(1, 1) - T(2, 2) + T(2, 1)) / NC
```

```
' Output Results
    Call TIMEOUT(JN, JR, SMC(), NV(), Z(), G(), T())
' Terminate
    End Sub

    Public Sub METHODHOLDERxBS(NC, JN, KN, JR, SMC(), DLB(), DUB(), NV(), NMC(), Z(), G(), C(),
T())
' A Subroutine to Invoke and Time an Operation or Method Elaboration
'    Arguments:-
'    NC    The Number of Cycles
'    JN    The Job Number
'    KN    The Constants' Set Number
'    JR    The Range Number
'    SMC() The Method Code
'    DLB() The Range Lower Bound
'    DUB() The Range Upper Bound
'    NV()  The Number of Intervals
'    NMC() The Number of Method Constants
'    Z()   The Array of Method Arguments
'    G()   The Array of Method Solutions
'    C()   The Array of Method Constants
'    T()   The Array of Times
'
' Set Local Constants
    NW = NV(JR)
    NX = NMC(KN)
    DD = DLB(JR)
    ZI = (DUB(JR) - DLB(JR)) / NV(JR)
' Do Composite
    T(1, 1) = Timer
    For I = 1 To NC
      For J = 0 To NW
        Z(J) = DD + J * ZI
        Call METHODxBS(KN, NX, Z(J), G(J), C())
      Next J
    Next I
    T(1, 2) = Timer
' Do Carcass
    T(2, 1) = Timer
    For I = 1 To NC
      For J = 0 To NW
        Z(J) = DD + J * ZI
        Call METHODxBScarc(KN, NX, Z(J), G(J), C())
      Next J
    Next I
    T(2, 2) = Timer
' Calculate Mill Time
    T(3, 1) = (T(1, 2) - T(1, 1) - T(2, 2) + T(2, 1)) / NC
' Output Results
    Call TIMEOUT(JN, JR, SMC(), NV(), Z(), G(), T())
' Terminate
    End Sub

    Public Sub METHODHOLDERxWS(NC, JN, KN, JR, SMC(), DLB(), DUB(), NV(), NMC(), Z(), G(), C(),
T())
' A Subroutine to Invoke and Time an Operation or Method Elaboration
'    Arguments:-
'    NC    The Number of Cycles
'    JN    The Job Number
'    KN    The Constants' Set Number
'    JR    The Range Number
'    SMC() The Method Code
'    DLB() The Range Lower Bound
'    DUB() The Range Upper Bound
```

```
'       NV()   The Number of Intervals
'       NMC()  The Number of Method Constants
'       Z()    The Array of Method Arguments
'       G()    The Array of Method Solutions
'       C()    The Array of Method Constants
'       T()    The Array of Times
'
' Set Local Constants
     NW = NV(JR)
     NX = NMC(KN)
     DD = DLB(JR)
     ZI = (DUB(JR) - DLB(JR)) / NV(JR)
' Do Composite
     T(1, 1) = Timer
     For I = 1 To NC
       For J = 0 To NW
         Z(J) = DD + J * ZI
         Call METHODxWS(KN, NX, Z(J), G(J), C())
       Next J
     Next I
     T(1, 2) = Timer
' Do Carcass
     T(2, 1) = Timer
     For I = 1 To NC
       For J = 0 To NW
         Z(J) = DD + J * ZI
         Call METHODxWScarc(KN, NX, Z(J), G(J), C())
       Next J
     Next I
     T(2, 2) = Timer
' Calculate Mill Time
     T(3, 1) = (T(1, 2) - T(1, 1) - T(2, 2) + T(2, 1)) / NC
' Output Results
     Call TIMEOUT(JN, JR, SMC(), NV(), Z(), G(), T())
' Terminate
     End Sub


     Public Sub METHODxDP(KN, NX, ZZ, GG, C())
' A Subroutine to Compute a Complete Gamma Function by means of
' The Davis Series, implemented as a polynomial Power Series
'     Arguments:-
'       KN     The Constants' Set Number
'       NX     The Number of Method Constants
'       ZZ     The Argument
'       GG     The Complete Gamma Function Solution
'       C()    The Array of Method Constants
'
     GG = 0#
     For I = 1 To NX
       GG = GG + C(KN, I) * ZZ ^ I
     Next I
     GG = 1 / GG
     End Sub


     Public Sub METHODxDR(KN, NX, ZZ, GG, C())
' A Subroutine to Compute a Complete Gamma Function by means of
' The Davis Series, implemented as a progressive Product Series
'     Arguments:-
'       KN     The Constants' Set Number
'       NX     The Number of Method Constants
'       ZZ     The Argument
'       GG     The Complete Gamma Function Solution
'       C()    The Array of Method Constants
'
```

```
      P = ZZ
      GG = 0#
      For I = 1 To NX
        GG = GG + P * C(KN, I)
        P = P * ZZ
      Next I
      GG = 1 / GG
      End Sub

      Public Sub METHODxES(KN, NX, ZZ, GG, C())
' A Subroutine to Compute a Complete Gamma Function by means of
' Exponentiated Summation
'     Arguments:-
'       KN     The Constants' Set Number
'       NX     The Number of Method Constants
'       ZZ     The Argument
'       GG     The Complete Gamma Function Solution
'       C()    The Array of Method Constants
'
' ( PI, PI2, HP, RC and EMC are
'   Common Shared by Public Declaration )
'
      DSUM = 0#
      For KK = 2 To 10
        COKK = KK
        DSUM = DSUM + ((-1) ^ KK * ZETA(COKK, 2000) * ZZ ^ KK) / KK
      Next KK
      GG = 1 / (ZZ * Exp(ZZ * EMC - DSUM))
      End Sub

      Public Sub METHODxEP(KN, NX, ZZ, GG, C())
' A Subroutine to Compute a Complete Gamma Function by means of
' The Euler Product
'     Arguments:-
'       KN     The Constants' Set Number
'       NX     The Number of Method Constants
'       ZZ     The Argument
'       GG     The Complete Gamma Function Solution
'       C()    The Array of Method Constants
'
' ( PI, PI2, HP, RC and EMC are
'   Common Shared by Public Declaration )
'
      PROD = 1#
      For I = 1 To 10
        PROD = PROD * (1 + ZZ / I) * Exp(-ZZ / I)
      Next I
      GG = 1 / (ZZ * Exp(ZZ * EMC) * PROD)
      End Sub

      Public Sub METHODxLA(KN, NX, ZZ, GG, C())
' A Subroutine to Compute a Complete Gamma Function by means of
' The Lanczos Approximation
'     Arguments:-
'       KN     The Constants' Set Number
'       NX     The Number of Method Constants
'       ZZ     The Argument
'       GG     The Complete Gamma Function Solution
'       C()    The Array of Method Constants
'
' ( PI, PI2, HP, RC and EMC are
'   Common Shared by Public Declaration )
'
      DSUM = 0#
```

```
    For I = 2 To 7
      DSUM = DSUM + C(KN, I) / (ZZ + I - 1)
    Next I
    GG = ((RC / ZZ) * (C(KN, 1) + DSUM)) * ((ZZ + 5.5) ^ (ZZ + 0.5)) * Exp(-(ZZ + 5.5))
    End Sub


    Public Sub METHODxHT(KN, NX, ZZ, GG, C())
' A Subroutine to Compute a Complete Gamma Function by means of
' The Hastings Type II ( Nine-Figure ) Polynomial
'    Arguments:-
'      KN    The Constants' Set Number
'      NX    The Number of Method Constants
'      ZZ    The Argument
'      GG    The Complete Gamma Function Solution
'      C()   The Array of Method Constants
'
' ( PI, PI2, HP, RC and EMC are
'   Common Shared by Public Declaration )
'
    DSUM = 0#
    For I = 1 To 9
      DSUM = DSUM + C(KN, I) * ZZ ^ (I - 1)
    Next I
    GG = DSUM / ZZ
    End Sub



    Public Sub METHODxST(KN, NX, ZZ, GG, C())
' A Subroutine to Compute a Complete Gamma Function by means of
' The Stirling Formula
'    Arguments:-
'      KN    The Constants' Set Number
'      NX    The Number of Method Constants
'      ZZ    The Argument
'      GG    The Complete Gamma Function Solution
'      C()   The Array of Method Constants
'
' ( PI, PI2, HP, RC and EMC are
'   Common Shared by Public Declaration )
'
    GG = RC * Exp(-ZZ) * ZZ ^ (ZZ - 0.5)
    End Sub

    Public Sub METHODxSE(KN, NX, ZZ, GG, C())
' A Subroutine to Compute a Complete Gamma Function by means of
' The Extended Stirling Formula
'    Arguments:-
'      KN    The Constants' Set Number
'      NX    The Number of Method Constants
'      ZZ    The Argument
'      GG    The Complete Gamma Function Solution
'      C()   The Array of Method Constants
'
' ( PI, PI2, HP, RC and EMC are
'   Common Shared by Public Declaration )
'
    DSUM = 1#
    For I = 2 To 8
      DSUM = DSUM + C(KN, I) / (ZZ ^ (I - 1))
    Next I
    GG = RC * Exp(-ZZ) * ZZ ^ (ZZ - 0.5) * DSUM
    End Sub

    Public Sub METHODxWO(KN, NX, ZZ, GG, C())
```

```
' A Subroutine to Compute a Complete Gamma Function by means of
' The Warren Type I Formula
'    Arguments:-
'      KN    The Constants' Set Number
'      NX    The Number of Method Constants
'      ZZ    The Argument
'      GG    The Complete Gamma Function Solution
'      C()   The Array of Method Constants
'
' ( PI, PI2, HP, RC and EMC are
'   Common Shared by Public Declaration )
'
     GG = RC * ZZ ^ (ZZ - 0.5) * Exp(-ZZ) * (1 + 1 / (8 * ZZ)) * (1 - 1 / (8 * PI * ZZ))
     End Sub

     Public Sub METHODxWT(KN, NX, ZZ, GG, C())
' A Subroutine to Compute a Complete Gamma Function by means of
' The Warren Type II Formula ( configured for Range One )
'    Arguments:-
'      KN    The Constants' Set Number
'      NX    The Number of Method Constants
'      ZZ    The Argument
'      GG    The Complete Gamma Function Solution
'      C()   The Array of Method Constants
'
' ( PI, PI2, HP, RC and EMC are
'   Common Shared by Public Declaration )
'
     DSUM = 0#
     For I = 0 To 3
       DSUM = DSUM + C(KN, I + 1) * (1 / ZZ) ^ I
     Next I
     GG = RC * ZZ ^ (ZZ - 0.5) * Exp(-ZZ) * (1 + 1 / (8 * ZZ)) * (1 - 1 / (8 * PI * ZZ)) * DSUM
     End Sub

     Public Sub METHODxBS(KN, NX, ZZ, GG, C())
' A Subroutine to Compute a Complete Gamma Function by means of
' The Burnside Formula
'    Arguments:-
'      KN    The Constants' Set Number
'      NX    The Number of Method Constants
'      ZZ    The Argument
'      GG    The Complete Gamma Function Solution
'      C()   The Array of Method Constants
'
' ( PI, PI2, HP, RC and EMC are
'   Common Shared by Public Declaration )
'
     GG = ((ZZ + 0.5) ^ (ZZ + 0.5) * Exp(-ZZ - 0.5) * RC) / ZZ
     End Sub

     Public Sub METHODxWS(KN, NX, ZZ, GG, C())
' A Subroutine to Compute a Complete Gamma Function by means of
' The Windshitl Formula
'    Arguments:-
'      KN    The Constants' Set Number
'      NX    The Number of Method Constants
'      ZZ    The Argument
'      GG    The Complete Gamma Function Solution
'      C()   The Array of Method Constants
'
' ( PI, PI2, HP, RC, EMC and ENAP are
'   Common Shared by Public Declaration )
TEST = DSINH(1)
```

```
    GG = Sqr(PI2 / ZZ) * ((ZZ / ENAP) + Sqr(ZZ * DSINH(1 / ZZ) * (1 / (810 * ZZ ^ 6)))) ^ ZZ
    End Sub

    Public Function DSINH(X)
' A Function to Return the Double-Precision Hyperbolic Sine for Argument X
'    Argument:-
'       X     The Double-Precision Argument of DSINH
'
    DSINH = (Exp(X) - Exp(-X)) / 2
    End Function

    Public Function ZETA(ZK, IZ)
' A Function to Return the Double-Precision Riemann's Zeta Function for Argument ZK.
' The Definitional Order(1/n) Algorithm is Employed.
'    Arguments:-
'       ZK    The Double-Precision Argument of ZETA
'       IZ    The Number of Iteration Cycles
'
    ZETA = 0#
    For K = 1 To IZ
     ZETA = ZETA + 1 / K ^ ZK
    Next K
    End Function

    Public Sub METHODxDPcarc(KN, NX, ZZ, GG, C())
' A Subroutine to Compute a Complete Gamma Function by means of
' The Davis Series, implemented as a polynomial Power Series
'    Arguments:-
'       KN    The Constants' Set Number
'       NX    The Number of Method Constants
'       ZZ    The Argument
'       GG    The Complete Gamma Function Solution
'       C()   The Array of Method Constants
'
    End Sub

    Public Sub METHODxDRcarc(KN, NX, ZZ, GG, C())
' A Subroutine to Compute a Complete Gamma Function by means of
' The Davis Series, implemented as a progressive Product Series
'    Arguments:-
'       KN    The Constants' Set Number
'       NX    The Number of Method Constants
'       ZZ    The Argument
'       GG    The Complete Gamma Function Solution
'       C()   The Array of Method Constants
'
    End Sub

    Public Sub METHODxEScarc(KN, NX, ZZ, GG, C())
' A Subroutine to Compute a Complete Gamma Function by means of
' Exponentiated Summation
'    Arguments:-
'       KN    The Constants' Set Number
'       NX    The Number of Method Constants
'       ZZ    The Argument
'       GG    The Complete Gamma Function Solution
'       C()   The Array of Method Constants
'
' ( PI, PI2, HP, RC and EMC are
'   Common Shared by Public Declaration )
'
    End Sub
```

```
        Public Sub METHODxEPcarc(KN, NX, ZZ, GG, C())
' A Subroutine to Compute a Complete Gamma Function by means of
' The Euler Product
'     Arguments:-
'       KN    The Constants' Set Number
'       NX    The Number of Method Constants
'       ZZ    The Argument
'       GG    The Complete Gamma Function Solution
'       C()   The Array of Method Constants
'
' ( PI, PI2, HP, RC and EMC are
'   Common Shared by Public Declaration )
'
        End Sub

        Public Sub METHODxLAcarc(KN, NX, ZZ, GG, C())
' A Subroutine to Compute a Complete Gamma Function by means of
' The Lanczos Approximation
'     Arguments:-
'       KN    The Constants' Set Number
'       NX    The Number of Method Constants
'       ZZ    The Argument
'       GG    The Complete Gamma Function Solution
'       C()   The Array of Method Constants
'
' ( PI, PI2, HP, RC and EMC are
'   Common Shared by Public Declaration )
'
        End Sub

        Public Sub METHODxHTcarc(KN, NX, ZZ, GG, C())
' A Subroutine to Compute a Complete Gamma Function by means of
' The Hastings Type II ( Nine-Figure ) Polynomial
'     Arguments:-
'       KN    The Constants' Set Number
'       NX    The Number of Method Constants
'       ZZ    The Argument
'       GG    The Complete Gamma Function Solution
'       C()   The Array of Method Constants
'
' ( PI, PI2, HP, RC and EMC are
'   Common Shared by Public Declaration )
'
        End Sub

        Public Sub METHODxSTcarc(KN, NX, ZZ, GG, C())
' A Subroutine to Compute a Complete Gamma Function by means of
' The Stirling Formula
'     Arguments:-
'       KN    The Constants' Set Number
'       NX    The Number of Method Constants
'       ZZ    The Argument
'       GG    The Complete Gamma Function Solution
'       C()   The Array of Method Constants
'
' ( PI, PI2, HP, RC and EMC are
'   Common Shared by Public Declaration )
'
        End Sub

        Public Sub METHODxSEcarc(KN, NX, ZZ, GG, C())
' A Subroutine to Compute a Complete Gamma Function by means of
' The Extended Stirling Formula
'     Arguments:-
```

```
'          KN    The Constants' Set Number
'          NX    The Number of Method Constants
'          ZZ    The Argument
'          GG    The Complete Gamma Function Solution
'          C()   The Array of Method Constants
'
' ( PI, PI2, HP, RC and EMC are
'   Common Shared by Public Declaration )
'
       End Sub

       Public Sub METHODxWOcarc(KN, NX, ZZ, GG, C())
' A Subroutine to Compute a Complete Gamma Function by means of
' The Warren Type I Formula
'     Arguments:-
'          KN    The Constants' Set Number
'          NX    The Number of Method Constants
'          ZZ    The Argument
'          GG    The Complete Gamma Function Solution
'          C()   The Array of Method Constants
'
' ( PI, PI2, HP, RC and EMC are
'   Common Shared by Public Declaration )
'
       End Sub

       Public Sub METHODxWTcarc(KN, NX, ZZ, GG, C())
' A Subroutine to Compute a Complete Gamma Function by means of
' The Warren Type II Formula ( configured for Range One )
'     Arguments:-
'          KN    The Constants' Set Number
'          NX    The Number of Method Constants
'          ZZ    The Argument
'          GG    The Complete Gamma Function Solution
'          C()   The Array of Method Constants
'
' ( PI, PI2, HP, RC and EMC are
'   Common Shared by Public Declaration )
'
       End Sub

       Public Sub METHODxBScarc(KN, NX, ZZ, GG, C())
' A Subroutine to Compute a Complete Gamma Function by means of
' The Burnside Formula
'     Arguments:-
'          KN    The Constants' Set Number
'          NX    The Number of Method Constants
'          ZZ    The Argument
'          GG    The Complete Gamma Function Solution
'          C()   The Array of Method Constants
'
' ( PI, PI2, HP, RC and EMC are
'   Common Shared by Public Declaration )
'
       End Sub

       Public Sub METHODxWScarc(KN, NX, ZZ, GG, C())
' A Subroutine to Compute a Complete Gamma Function by means of
' The Windshitl Formula
'     Arguments:-
'          KN    The Constants' Set Number
'          NX    The Number of Method Constants
'          ZZ    The Argument
'          GG    The Complete Gamma Function Solution
```

```
'     C()   The Array of Method Constants
'
' ( PI, PI2, HP, RC and EMC are
'   Common Shared by Public Declaration )

     End Sub
```

**The Methods Accuracy-Speed
Data Summary Tabulation**

Points per Range		101			

Mill Times are in Microseconds

Method Code	ES	EP	DP	LA	
Number of Method Applications		2	2	3	3
INTE1000		9482.871287	6.485148515	16.93069307	2.508250825
INTE2000		9467.871287	6.658415842	16.33663366	3.085808581
INTE3500		9565.82744	6.591230552	16.99198491	3.064592174
INTE3500A		9498.076379	6.775106082	17.02970297	3.064592174
INTE5000		9511.019802	6.574257426	16.85148515	3.069306931
Mean		9505.133239	6.616831683	16.82809995	2.958510137
Population Standard Deviation		33.62226785	0.096528203	0.253027028	0.225264763
Percentage Coefficient of Variation		0.353727475	1.458828144	1.503598321	7.614128498
$\log_e$(Mean)		9.159587273	1.889616658	2.823050105	1.084685809
$\log_e$(Population SD)		3.515188581	-2.337920056	-1.374258965	-1.490478842

Log_{128} Statistics

Points per Range		101			

Mill Times are in Microseconds

Method Code	ES	EP	DP	LA	
Number of Method Applications		2	2	3	3
INTE1000		1.887301177	0.385305659	0.583081304	0.189525946
INTE2000		1.886974911	0.390739854	0.575719832	0.23223551
INTE3500		1.889096293	0.388649691	0.583826069	0.230813585
INTE3500A		1.887631377	0.394320505	0.584283052	0.230813585
INTE5000		1.887912046	0.388118278	0.582114834	0.231130418
Mean		1.887784448	0.389448654	0.581828627	0.223552977
Population Standard Deviation		0.724477876	-0.481843667	-0.283233799	-0.307186633
Percentage Coefficient of Variation		-0.214184259	0.077829991	0.084059887	0.418382702

Plotting Selection

Method Code	ES	EP	DP	LA	
$\mu+\sigma$		2.612262324	-0.092395013	0.298594828	-0.083633657
$\mu-\sigma$		1.163306572	0.871292322	0.865062426	0.53073961
μ		1.887784448	0.389448654	0.581828627	0.223552977

Accuracy Summary ex ACCSTATS

Code	ES	EP	DP	LA	
Method	Exponentiated Summation	Euler Product	Davis Series	Lanczos	
n	101	101	101	101	
$\Sigma\delta 2$	1.0032E+57	5.2048E-01	1.2851E+15	2.3776E+13	
RMS	2.2285E+27	6.1623E-02	2.0594E+06	2.8013E+05	
s	9.9694E+55	3.7619E-03	6.4258E+13	1.1879E+12	
σ	9.9199E+55	3.7432E-03	6.3940E+13	1.1820E+12	
$\sigma/(n)0.5$	9.8707E+54	3.7246E-04	6.3622E+12	1.1761E+11	
MEF	-1.2147E+01	1.4073E+00	6.3936E+00	1.2042E+00	
MAX(ds)	2.2382E+30	-2.2547E+00	5.3142E-09	-3.9292E+00	
MIN(ds)	2.4194E+00	-1.0462E+01	-3.3333E+01	-8.5511E+00	

Log_{128}(statistics)

n	9.5117E-01	9.5117E-01	9.5117E-01	9.5117E-01	
$\Sigma\delta 2$	2.7051E+01	-1.3459E-01	7.1701E+00	6.3478E+00	
RMS	1.2978E+01	-5.7434E-01	2.9963E+00	2.5851E+00	
s	2.6575E+01	-1.1506E+00	6.5527E+00	5.7302E+00	
σ	2.6574E+01	-1.1516E+00	6.5517E+00	5.7292E+00	
$\sigma/(n)0.5$	2.6098E+01	-1.6272E+00	6.0761E+00	5.2536E+00	
MAX(ds)	1.4403E+01	-1.6756E-01	-3.9268E+00	-2.8203E-01	
MIN(ds)	1.8210E-01	-4.8387E-01	-7.2270E-01	-4.4230E-01	

Digit Frequency

Method Code	ES	EP	DP	LA	
Mean Point Mill Time (seconds)	0.009505133	6.61683E-06	1.68281E-05	2.95851E-06	
Mean Equivalent Figure	-1.2147E+01	1.4073E+00	6.3936E+00	1.2042E+00	
Digit Frequency	-1.2780E+03	2.1269E+05	3.7993E+05	4.0704E+05	

		IIT	ST	SE	WO
Points per Range		101			
Mill Times are in Microseconds					
Method Code		IIT	ST	SE	WO
Number of Method Applications		1	1	1	1
INTE1000		5.544554455	1.188118812	5.544554455	0.99009901
INTE2000		5.99009901	1.089108911	5.693069307	1.386138614
INTE3500		5.742574257	0.90523338	5.601131542	1.244695898
INTE3500A		5.912305516	0.933521924	5.601131542	1.244695898
INTE5000		5.96039604	0.871287129	5.782178218	1.207920792
Mean		5.829985856	0.997454031	5.644413013	1.214710042
Population Standard Deviation		0.166452057	0.121048524	0.083743644	0.127814137
Percentage Coefficient of Variation		2.855102245	12.13574963	1.4836555	10.52219316
Log_e(Mean)		1.763014574	-0.002549215	1.730666209	0.1945054
Log_e(Population SD)		-1.793047957	-2.11156379	-2.479995003	-2.057178125

Log_{128} Statistics

		IIT	ST	SE	WO
Points per Range		101			
Mill Times are in Microseconds					
Method Code		IIT	ST	SE	WO
Number of Method Applications		1	1	1	1
INTE1000		0.353010219	0.035525588	0.353010219	-0.002050756
INTE2000		0.368939978	0.017592604	0.358458095	0.067295933
INTE3500		0.360242515	-0.020519759	0.355102616	0.04511333
INTE3500A		0.366245832	-0.014177742	0.355102616	0.04511333
INTE5000		0.367915456	-0.028397123	0.361659011	0.038932265
Mean		0.363356055	-0.000525391	0.35668908	0.040087425
Population Standard Deviation		-0.369545914	-0.435191801	-0.511125213	-0.423982954
Percentage Coefficient of Variation		0.216220345	0.514455903	0.08130802	0.485051933

Plotting Selection

Method Code	HT	ST	SE	WO
$\mu+\sigma$	-0.006189859	-0.435717193	-0.154436134	-0.383895529
$\mu-\sigma$	0.732901968	0.43466641	0.867814293	0.464070379
μ	0.363356055	-0.000525391	0.35668908	0.040087425

Accuracy Summary ex ACCSTATS

Code	HT	ST	SE	WO
Method	Hastings Type II	Stirling Formula	Extended Stirling Formula	Warren Type I
n	101	101	101	101
$\Sigma\delta 2$	9.2361E-03	1.8965E+11	2.0814E-04	3.7993E+07
RMS	9.5628E-03	4.3332E+04	1.4355E-03	6.1333E+02
s	2.3937E-04	9.3137E+09	1.0313E-05	1.8859E+06
σ	2.3818E-04	9.2675E+09	1.0262E-05	1.8765E+06
$\sigma/(n)0.5$	2.3700E-05	9.2215E+08	1.0211E-06	1.8672E+05
MEF	2.3693E+00	1.7287E+00	5.1862E+00	3.7661E+00
MAX(ds)	2.1652E-05	-6.9188E-01	2.8676E-02	1.0737E-02
MIN(ds)	-3.5868E+00	-7.7863E+00	2.3085E-08	-3.8729E-01

Log_{128}(statistics)

	HT	ST	SE	WO
n	9.5117E-01	9.5117E-01	9.5117E-01	9.5117E-01
$\Sigma\delta 2$	-9.6550E-01	5.3521E+00	-1.7472E+00	3.5970E+00
RMS	-9.5834E-01	2.2005E+00	-1.3492E+00	1.3229E+00
s	-1.7184E+00	4.7310E+00	-2.3665E+00	2.9781E+00
σ	-1.7194E+00	4.7299E+00	-2.3675E+00	2.9771E+00
$\sigma/(n)0.5$	-2.1950E+00	4.2543E+00	-2.8431E+00	2.5015E+00
MAX(ds)	-2.2136E+00	7.5915E-02	-7.3200E-01	-9.3447E-01
MIN(ds)	-2.6324E-01	-4.2299E-01	-3.6241E+00	1.9550E-01

Digit Frequency

Method Code	HT	ST	SE	WO
Mean Point Mill Time (seconds)	5.82999E-06	9.97454E-07	5.64441E-06	1.21471E-06
Mean Equivalent Figure	2.3693E+00	1.7287E+00	5.1862E+00	3.7661E+00
Digit Frequency	4.0640E+05	1.7331E+06	9.1883E+05	3.1004E+06

Points per Range		101			
Mill Times are in Microseconds					
Method Code	WT		BS	WS	
Number of Method Applications		1	1	1	

	WT	BS	WS
INTE1000	4.257425743	0.99009901	3.861386139
INTE2000	3.762376238	1.089108911	3.514851485
INTE3500	4.186704385	1.103253182	3.394625177
INTE3500A	4.186704385	1.074964639	3.42291372
INTE5000	4.118811881	0.99009901	3.485148515
Mean	4.102404526	1.04950495	3.535785007
Population Standard Deviation	0.175575026	0.04932276	0.16833561
Percentage Coefficient of Variation	4.27980772	4.699621506	4.760911935
Log_e(Mean)	1.411573272	0.048318577	1.262935342
Log_e(Population SD)	-1.739688831	-3.009369634	-1.781795611

Log_{128} Statistics

Points per Range		101			
Mill Times are in Microseconds					
Method Code	WT		BS	WS	
Number of Method Applications		1	1	1	

	WT	BS	WS
INTE1000	0.298568767	-0.002050756	0.278445547
INTE2000	0.273092018	0.017592604	0.259066247
INTE3500	0.295116437	0.020251987	0.251893184
INTE3500A	0.295116437	0.014898458	0.253603561
INTE5000	0.291746891	-0.002050756	0.257317162
Mean	0.290924251	0.009958425	0.260290079
Population Standard Deviation	-0.358548636	-0.62022895	-0.367226813
Percentage Coefficient of Variation	0.299649426	0.318934939	0.32160542

Plotting Selection

Method Code	WT	BS	WS
$\mu+\sigma$	-0.067624384	-0.610270525	-0.106936734
$\mu-\sigma$	0.649472887	0.630187374	0.627516892
μ	0.290924251	0.009958425	0.260290079

Accuracy Summary ex ACCSTATS

Code	WT	BS	WS
Method	Warren Type II	Burnside	Windshitl
n	101	101	101
$\Sigma\delta2$	3.0443E+08	4.4089E+10	1.8656E+11
RMS	1.7361E+03	2.0893E+04	4.2978E+04
s	1.4936E+07	2.1667E+09	9.1650E+09
σ	1.4861E+07	2.1559E+09	9.1195E+09
$\sigma/(n)0.5$	1.4788E+06	2.1452E+08	9.0743E+08
MEF	3.0123E+00	2.0912E+00	1.8553E+00
MAX(ds)	7.5306E-01	2.7508E+00	1.7615E+00
MIN(ds)	2.7699E-02	3.3376E-01	-2.8856E+00

Log_{128}(statistics)

n	9.5117E-01	9.5117E-01	9.5117E-01
$\Sigma\delta2$	4.0259E+00	5.0514E+00	5.3487E+00
RMS	1.5374E+00	2.0501E+00	2.1988E+00
s	3.4046E+00	4.4304E+00	4.7276E+00
σ	3.4036E+00	4.4294E+00	4.7266E+00
$\sigma/(n)0.5$	2.9280E+00	3.9538E+00	4.2510E+00
MAX(ds)	-5.8453E-02	2.0855E-01	1.1669E-01
MIN(ds)	-7.3915E-01	-2.2616E-01	-2.1841E-01

Digit Frequency

Method Code	WT	BS	WS
Mean Point Mill Time (seconds)	4.1024E-06	1.0495E-06	3.53579E-06
Mean Equivalent Figure	3.0123E+00	2.0912E+00	1.8553E+00
Digit Frequency	7.3429E+05	1.9925E+06	5.2472E+05

APPENDIX E

The Operations and Functions Timing Program Project TARIFF.vbp

```
Option Explicit

Private Declare Sub keybd_event Lib "user32" (ByVal bVk As Byte, ByVal _
    bScan As Byte, ByVal dwFlags As Long, ByVal dwExtraInfo As Long)

Private Declare Function GetVersionExA Lib "kernel32" _
    (lpVersionInformation As OSVERSIONINFO) As Integer

Private Type OSVERSIONINFO
    dwOSVersionInfoSize As Long
    dwMajorVersion As Long
    dwMinorVersion As Long
    dwBuildNumber As Long
    dwPlatformId As Long
    szCSDVersion As String * 128
End Type

Private Const KEYEVENTF_KEYUP = &H2
Private Const VK_SNAPSHOT = &H2C
Private Const VK_MENU = &H12

Dim blnAboveVer4 As Boolean

    Private Function CaptureScreen()
    If blnAboveVer4 Then
        keybd_event VK_SNAPSHOT, 0, 0, 0
      Else
        keybd_event VK_SNAPSHOT, 1, 0, 0
    End If
    End Function

    Private Function CaptureForm()
    If blnAboveVer4 Then
        keybd_event VK_SNAPSHOT, 1, 0, 0
      Else
        keybd_event VK_MENU, 0, 0, 0
        keybd_event VK_SNAPSHOT, 0, 0, 0
        keybd_event VK_SNAPSHOT, 0, KEYEVENTF_KEYUP, 0
        keybd_event VK_MENU, 0, KEYEVENTF_KEYUP, 0
    End If
    End Function

    Private Sub SaveBitMap(TariInputFormFileName)
' Load the captured image into a PictureBox and print it
    InputPicture.Picture = Clipboard.GetData()
    SavePicture InputPicture.Picture, TariInputFormFileName + ".bmp"
' Clipboard.Clear
'   Picture1.Picture = Clipboard.GetData()
'   Printer.PaintPicture Picture1.Picture, 0, 0
'   Printer.EndDoc
    End Sub


    Private Sub Exit_Click()
    Unload TariOutputForm
    Unload TariServiceForm
    Unload TariPlotForm
    Unload Me
    End Sub

    Private Sub Form_Load()
    Dim osinfo As OSVERSIONINFO
    Dim retvalue As Integer
    osinfo.dwOSVersionInfoSize = 148
```

```
osinfo.szCSDVersion = Space$(128)
retvalue = GetVersionExA(osinfo)
If osinfo.dwMajorVersion > 4 Then blnAboveVer4 = True
InputPicture.Visible = True
End Sub

Private Sub Compute_Click()
Call DoCompute
End Sub
```

```
Option Explicit

Private Declare Sub keybd_event Lib "user32" (ByVal bVk As Byte, ByVal _
    bScan As Byte, ByVal dwFlags As Long, ByVal dwExtraInfo As Long)

Private Declare Function GetVersionExA Lib "kernel32" _
    (lpVersionInformation As OSVERSIONINFO) As Integer

Private Type OSVERSIONINFO
    dwOSVersionInfoSize As Long
    dwMajorVersion As Long
    dwMinorVersion As Long
    dwBuildNumber As Long
    dwPlatformId As Long
    szCSDVersion As String * 128
End Type

Private Const KEYEVENTF_KEYUP = &H2
Private Const VK_SNAPSHOT = &H2C
Private Const VK_MENU = &H12

Dim blnAboveVer4 As Boolean

    Private Function CaptureScreen()
    If blnAboveVer4 Then
        keybd_event VK_SNAPSHOT, 0, 0, 0
     Else
        keybd_event VK_SNAPSHOT, 1, 0, 0
    End If
    End Function

    Private Function CaptureForm()
    If blnAboveVer4 Then
        keybd_event VK_SNAPSHOT, 1, 0, 0
     Else
        keybd_event VK_MENU, 0, 0, 0
        keybd_event VK_SNAPSHOT, 0, 0, 0
        keybd_event VK_SNAPSHOT, 0, KEYEVENTF_KEYUP, 0
        keybd_event VK_MENU, 0, KEYEVENTF_KEYUP, 0
    End If
    End Function

    Private Sub SaveBitMap(TariOutputFormFileName)
    Dim OutputPicture As Image
    OutputPicture.Picture = Clipboard.GetData()
    SavePicture OutputPicture.Picture, TariOutputFormFileName + ".bmp"
'    SavePicture Clipboard.GetData(), TariOutputFormFileName + ".bmp"
    End Sub

    Private Sub Form_Load()
    Dim osinfo As OSVERSIONINFO
    Dim retvalue As Integer
    osinfo.dwOSVersionInfoSize = 148
    osinfo.szCSDVersion = Space$(128)
    retvalue = GetVersionExA(osinfo)
    If osinfo.dwMajorVersion > 4 Then blnAboveVer4 = True
'    OutputPicture.Visible = True
    End Sub
```

```
Option Explicit

Private Declare Sub keybd_event Lib "user32" (ByVal bVk As Byte, ByVal _
    bScan As Byte, ByVal dwFlags As Long, ByVal dwExtraInfo As Long)

Private Declare Function GetVersionExA Lib "kernel32" _
    (lpVersionInformation As OSVERSIONINFO) As Integer

Private Type OSVERSIONINFO
    dwOSVersionInfoSize As Long
    dwMajorVersion As Long
    dwMinorVersion As Long
    dwBuildNumber As Long
    dwPlatformId As Long
    szCSDVersion As String * 128
End Type

Private Const KEYEVENTF_KEYUP = &H2
Private Const VK_SNAPSHOT = &H2C
Private Const VK_MENU = &H12

Dim blnAboveVer4 As Boolean

Private Function CaptureScreen()
    If blnAboveVer4 Then
        keybd_event VK_SNAPSHOT, 0, 0, 0
    Else
        keybd_event VK_SNAPSHOT, 1, 0, 0
    End If
End Function

Private Function CaptureForm()
    If blnAboveVer4 Then
        keybd_event VK_SNAPSHOT, 1, 0, 0
    Else
        keybd_event VK_MENU, 0, 0, 0
        keybd_event VK_SNAPSHOT, 0, 0, 0
        keybd_event VK_SNAPSHOT, 0, KEYEVENTF_KEYUP, 0
        keybd_event VK_MENU, 0, KEYEVENTF_KEYUP, 0
    End If
End Function

Private Sub SaveBitMap(TariInputFormFileName)
' Load the captured image into a PictureBox and print it
    Picture1.Picture = Clipboard.GetData()
    SavePicture Picture1.Picture, TariInputFormFileName + ".bmp"
'   Picture1.Picture = Clipboard.GetData()
'   Printer.PaintPicture Picture1.Picture, 0, 0
'   Printer.EndDoc
End Sub

    Private Sub Form_Load()
    Dim osinfo As OSVERSIONINFO
    Dim retvalue As Integer
    osinfo.dwOSVersionInfoSize = 148
    osinfo.szCSDVersion = Space$(128)
    retvalue = GetVersionExA(osinfo)
    If osinfo.dwMajorVersion > 4 Then blnAboveVer4 = True
    Picture1.Visible = True
    End Sub

    Private Sub Quit_Click()
' A Subroutine to Unload the Project
    TariInputForm.Hide
```

```
        TariOutputForm.Hide
        TariServiceForm.Hide
        TariPlotForm.Hide
        Unload Me
        End Sub

        Private Sub Size_DblClick()
' A Subroutine to Adjust the Forms' Size
        Dim MP As Integer
        MP = TariServiceForm.Size
        Cantonise (MP)
        End Sub

        Private Sub TariInputFormPrint_Click()
' A Subroutine to Print the Input Form
        TariInputForm.PrintForm
        End Sub

        Private Sub TariOutputFormPrint_Click()
' A Subroutine to Print the Output Form
        TariOutputForm.PrintForm
        End Sub

        Private Sub TariPlotFormPrint_Click()
' A Subroutine to Print the Plot Form
        TariPlotForm.PrintForm
        End Sub

        Private Sub TariInputFormSave_Click()
' A Subroutine to Save a Bitmap Image of the Input Form
        DoEvents
        TariInputForm.SetFocus
        Call CaptureForm
        DoEvents
        Call CaptureForm
        DoEvents
        Call SaveBitMap(TariInputFormFileName)
        TariServiceForm.SetFocus
        End Sub

        Private Sub TariOutputFormSave_Click()
' A Subroutine to Save the Output Form
        DoEvents
        TariOutputForm.SetFocus
        Call CaptureForm
        DoEvents
        Call CaptureForm
        DoEvents
        Picture1.Picture = Clipboard.GetData()
        SavePicture Picture1.Picture, TariOutputFormFileName + ".bmp"
        TariServiceForm.SetFocus
        End Sub

        Private Sub TariPlotFormSave_Click()
' A Subroutine to Save the Plot Form
        SavePicture TariPlotForm.Image, TariPlotFormFileName + ".bmp"
        End Sub
```

```
' VARIABLE TYPE DEFAULTS
    DefDbl A-H, O-R, T-Z
    DefStr S
    DefInt I-K, M-N
    DefLng L

    Public Sub DoCompute()
' Assign Service Form Product Bitmap Filenames
    TariServiceForm.TariInputFormFileName = TariInputForm.FileNameOut + "In"
    TariServiceForm.TariOutputFormFileName = TariInputForm.FileNameOut + "Out"
    TariServiceForm.TariPlotFormFileName = TariInputForm.FileNameOut + "Plot"
' Assign the Sequential Files' Transput Parameters
'   File One ( The Operations' Descriptors Input File )
    ILogicUnit(1) = 1
    STransmode(1) = "Input"
    SPath(1) = TariInputForm.Path
    SFile(1) = TariInputForm.FileNameIn
    SExt(1) = ".ODI"
'   File Two ( The Operations Timings Output File )
    ILogicUnit(2) = 2
    STransmode(2) = "Output"
    SPath(2) = TariInputForm.Path
    SFile(2) = TariInputForm.FileNameOut
    SExt(2) = ".OTO"
' Set Number of Cycles
    LNC = TariInputForm.LNC
' Define Random Number Range Parameters
    IRG(1, 1) = 1: IRG(1, 2) = 30: IRG(2, 1) = 1: IRG(2, 2) = 30
    DRG(1, 1) = 0: DRG(1, 2) = 1: DRG(2, 1) = 0.1: DRG(2, 2) = 10
    DRG(3, 1) = 0.1: DRG(3, 2) = 100: DRG(4, 1) = 0: DRG(4, 2) = -1000
' Load Operations Descriptions
    Call OPSIN(NOP, KO(), SOC(), SOP())
' Do Timings
    DStart = Timer
' ** Job 1 Block ( The Real Power of a Real Number )
    KN = 1
    TariOutputForm.Status = SOP(KN) + " BEGUN"
    Call OPHOLDERxPWR(KN, LNC, KO(), SOC(), SOP(), LTM())
    TariOutputForm.Status = SOP(KN) + " FINISHED"
' ** Job 2 Block ( The Real Square of a Real Number )
    KN = 2
    TariOutputForm.Status = SOP(KN) + " BEGUN"
    Call OPHOLDERxSQR(KN, LNC, KO(), SOC(), SOP(), LTM())
    TariOutputForm.Status = SOP(KN) + " FINISHED"
' ** Job 3 Block ( The Product of Two Real Numbers )
    KN = 3
    TariOutputForm.Status = SOP(KN) + " BEGUN"
    Call OPHOLDERxMUL(KN, LNC, KO(), SOC(), SOP(), LTM())
    TariOutputForm.Status = SOP(KN) + " FINISHED"
' ** Job 4 Block ( The Dividend of Two Real Numbers )
    KN = 4
    TariOutputForm.Status = SOP(KN) + " BEGUN"
    Call OPHOLDERxDIV(KN, LNC, KO(), SOC(), SOP(), LTM())
    TariOutputForm.Status = SOP(KN) + " FINISHED"
' ** Job 5 Block ( The Sum of Two Real Numbers )
    KN = 5
    TariOutputForm.Status = SOP(KN) + " BEGUN"
    Call OPHOLDERxADD(KN, LNC, KO(), SOC(), SOP(), LTM())
    TariOutputForm.Status = SOP(KN) + " FINISHED"
' ** Job 6 Block ( The Difference of Two Real Numbers )
    KN = 6
    TariOutputForm.Status = SOP(KN) + " BEGUN"
    Call OPHOLDERxSUB(KN, LNC, KO(), SOC(), SOP(), LTM())
    TariOutputForm.Status = SOP(KN) + " FINISHED"
```

```
' ** Job 7 Block ( The Assignment of a Real Number )
    KN = 7
    TariOutputForm.Status = SOP(KN) + " BEGUN"
    Call OPHOLDERxASS(KN, LNC, KO(), SOC(), SOP(), LTM())
    TariOutputForm.Status = SOP(KN) + " FINISHED"
' ** Job 8 Block ( The Bracketing of a Real Number )
    KN = 8
    TariOutputForm.Status = SOP(KN) + " BEGUN"
    Call OPHOLDERxBRT(KN, LNC, KO(), SOC(), SOP(), LTM())
    TariOutputForm.Status = SOP(KN) + " FINISHED"
' ** Job 9 Block ( The Napierian Exponent of a Real Number )
    KN = 9
    TariOutputForm.Status = SOP(KN) + " BEGUN"
    Call OPHOLDERxEXP(KN, LNC, KO(), SOC(), SOP(), LTM())
    TariOutputForm.Status = SOP(KN) + " FINISHED"
' ** Job 10 Block ( The Napierian Logarithm of a Real Number )
    KN = 10
    TariOutputForm.Status = SOP(KN) + " BEGUN"
    Call OPHOLDERxLGN(KN, LNC, KO(), SOC(), SOP(), LTM())
    TariOutputForm.Status = SOP(KN) + " FINISHED"
' ** Job 11 Block ( The Radix Ten Logarithm of a Real Number )
    KN = 11
    TariOutputForm.Status = SOP(KN) + " BEGUN"
    Call OPHOLDERxL10(KN, LNC, KO(), SOC(), SOP(), LTM())
    TariOutputForm.Status = SOP(KN) + " FINISHED"
' ** Job 12 Block ( The Zeta Function of a Real Number )
    KN = 12
    TariOutputForm.Status = SOP(KN) + " BEGUN"
    Call OPHOLDERxZET(KN, LNC, KO(), SOC(), SOP(), LTM())
    TariOutputForm.Status = SOP(KN) + " FINISHED"
' ** Job 13 Block ( The Hyperbolic Sine of a Real Number )
    KN = 13
    TariOutputForm.Status = SOP(KN) + " BEGUN"
    Call OPHOLDERxSNH(KN, LNC, KO(), SOC(), SOP(), LTM())
    TariOutputForm.Status = SOP(KN) + " FINISHED"
' ** Job 14 Block ( The Square Root of a Positive Real Number )
    KN = 14
    TariOutputForm.Status = SOP(KN) + " BEGUN"
    Call OPHOLDERxSRT(KN, LNC, KO(), SOC(), SOP(), LTM())
    TariOutputForm.Status = SOP(KN) + " FINISHED"
' ** Job 15 Block ( The Inverse Tangent of a Real Number )
    KN = 15
    TariOutputForm.Status = SOP(KN) + " BEGUN"
    Call OPHOLDERxATN(KN, LNC, KO(), SOC(), SOP(), LTM())
    TariOutputForm.Status = SOP(KN) + " FINISHED"
' ** Job 16 Block ( The Cosine of a Real Number )
    KN = 16
    TariOutputForm.Status = SOP(KN) + " BEGUN"
    Call OPHOLDERxCOS(KN, LNC, KO(), SOC(), SOP(), LTM())
    TariOutputForm.Status = SOP(KN) + " FINISHED"
' ** Job 17 Block ( The Sine of a Real Number )
    KN = 17
    TariOutputForm.Status = SOP(KN) + " BEGUN"
    Call OPHOLDERxSIN(KN, LNC, KO(), SOC(), SOP(), LTM())
    TariOutputForm.Status = SOP(KN) + " FINISHED"
' ** Job 18 Block ( FOR-NEXT Loop Access )
    KN = 18
    TariOutputForm.Status = SOP(KN) + " BEGUN"
    Call OPHOLDERxLAC(KN, LNC, KO(), SOC(), SOP(), LTM())
    TariOutputForm.Status = SOP(KN) + " FINISHED"
' ** Job 19 Block ( The Sine of a Real Number )
    KN = 19
    TariOutputForm.Status = SOP(KN) + " BEGUN"
    Call OPHOLDERxAR1(KN, LNC, KO(), SOC(), SOP(), LTM())
```

```
        TariOutputForm.Status = SOP(KN) + " FINISHED"
' ** Job 20 Block ( FOR-NEXT Loop Access )
     KN = 20
     TariOutputForm.Status = SOP(KN) + " BEGUN"
     Call OPHOLDERxAR2(KN, LNC, KO(), SOC(), SOP(), LTM())
     TariOutputForm.Status = SOP(KN) + " FINISHED"
' ** End of Job Blocks
' Output the Results
     Call OPSOUT(LNC, NOP, KO(), SOC(), SOP(), LTM())
     DFinish = Timer
     TariOutputForm.PMT = DFinish - DStart
' Terminate
     TariOutputForm.Status = "ALL FINISHED"
     Close
     End Sub
```

```
' VARIABLE TYPE DEFAULTS
    DefDbl A-H, O-R, T-Z
    DefStr S
    DefInt I-K, M-N
    DefLng L
' Declare the Timing Function GetTickCount()
    Private Declare Function GetTickCount Lib "kernel32" () As Long
' Object Definitions
'    ( None )
' Publicise Object Instantiations
'    ( None )
' Static Array Declarations
'    ( None )
' Pseudo-Static Array Declarations
    Public ILogicUnit(), STransmode(), SPath(), SFile(), SExt()
    Public IRG(), DRG(), AR1(), AR2()
    Public KO(), SOC(), SOP(), LTM()
' Dynamic Array Declarations
'    ( None )
' Constant Definitions
    Const PI As Double = 3.14159265358979
    Const PI2 As Double = 6.28318530717959
    Const HP As Double = 1.5707963267949
    Const RC As Double = 2.506628274631
    Const EMC As Double = 0.577215664901533
    Const ENAP As Double = 2.71828182845905
    Const L10 As Double = 2.30258509299405
' Declare Logical Unit Number Holders
    Public IU, IV, IW, IX
' Declare String Constant Data ( for Common Shared emulation )
    Public SC, SM, SCR
' Declare Formats ( for Common Shared emulation )
    Public SI4, SF10P8, SF11P6
' Declare Dynamic Array Dimensions
    Public NUnits
    Public NRX, NRY, NAR, NP
' Declare Plotting Parameters
'    ( None )

    Sub Main()
'    Program TARIFF
' A Program to Cantonise Input, Output, Service and Plot Forms; and
' To Compute a Set of Empirical Operation and Function Mill Times
' Based upon Data Located from TariInputForm
'    Written by:-
'
'        James R Warren BSc MSc PhD PGCE
'        "Southgate"
'        31 Victoria Avenue
'        Bloxwich
'        Walsall
'        WS3 3HS
'
'        5 September 2004
'
'    This Program is Written in MicroSoft Visual Basic 6.0
'
' Initialise the Random Number Generator
    Randomize (Timer)
' Assign the Pseudo-Static Array Dimensions Maxima and Dimension the Pseudo-Static Arrays
    NUnits = 10
    ReDim ILogicUnit(NUnits), STransmode(NUnits), SPath(NUnits), SFile(NUnits), SExt(NUnits)
    NRX = 4: NRY = 2
    ReDim IRG(NRX, NRY), DRG(NRX, NRY)
```

```
        NAR = 30
        ReDim AR1(NAR), AR2(NAR, NAR)
        NP = 20
        ReDim KO(NP), SOC(NP), SOP(NP), LTM(NP)
'       IModes = 3: ITimes = 2
'       ReDim T(IModes, ITimes)
' Assign the Dynamic Arrays' Maxima and Dimension the Dynamic Arrays
'       ( None )
' Form Definitions
        Dim TariInputForm As Form
        Dim TariOutputForm As Form
        Dim TariServiceForm As Form
        Dim TariPlotForm As Form
' String Constant Definitions
        SC = ":": SM = ",": SCR = Chr(13) + Chr(10)
' Format Definitions
        SI4 = "####": SF10P8 = "#.00000000": SF11P6 = "####.000000"
' Set MarginPercent
        MP = 12
' Perform Actions
        Call Cantonise(MP)
' Terminate
        End Sub


        Public Sub Cantonise(MP)
' A Subroutine to Set Up Window's Cantonisation and Display; and to
' Correct Excessively Small Size Choices
'       Argument:-
'         MP    The Margin Percent Windows' Scale Factor
'
        If MP > 40 Then MP = 40
' Define Variables
        Dim InsetOne, InsetTwo, DownsetOne, DownsetTwo As Integer
        Const SQRT2 As Double = 1.4142135623731
' Define Form Dimensions
        TariInputForm.Height = Screen.Height * (50 - MP) / 100
        TariInputForm.Width = SQRT2 * TariInputForm.Height
        TariOutputForm.Height = TariInputForm.Height
        TariServiceForm.Height = TariInputForm.Height
        TariPlotForm.Height = TariInputForm.Height
        TariOutputForm.Width = TariInputForm.Width
        TariServiceForm.Width = TariInputForm.Width
        TariPlotForm.Width = TariInputForm.Width
' Define Form Positions
        InsetOne = (Screen.Width - 2 * TariInputForm.Width) / 2
        InsetTwo = InsetOne + TariInputForm.Width
        DownsetOne = MP * Screen.Height / 100
        DownsetTwo = DownsetOne + TariInputForm.Height
        TariInputForm.Left = InsetOne
        TariInputForm.Top = DownsetOne
        TariOutputForm.Left = InsetTwo
        TariOutputForm.Top = DownsetOne
        TariServiceForm.Left = InsetOne
        TariServiceForm.Top = DownsetTwo
        TariPlotForm.Left = InsetTwo
        TariPlotForm.Top = DownsetTwo
' Show the Forms
        TariInputForm.Show
        TariOutputForm.Show
        TariServiceForm.Show
        TariPlotForm.Show
        End Sub

        Public Sub FileOpener(IFS, ILU)
```

```
' A Subroutine to Open a Sequential Access File whose Transput Properties
' Are Stored in Public ( Shared ) Arrays
'    Arguments:-
'       IFS    The File Execution Serial Number
'       ILU    The File Logical Unit Number
' ( Arrays ILogicUnit, STransMode, SPath, SFile and SExt
'   are Common Shared by static Public Declaration )
'
    ILU = ILogicUnit(IFS)
    SP = SPath(IFS) + SFile(IFS) + SExt(IFS)
    STM = STransmode(IFS)
    Select Case STM
      Case "Input"
        Open SP For Input As #ILU
      Case "Output"
        Open SP For Output As #ILU
      Case "Append"
        Open SP For Append As #ILU
    End Select
    End Sub


    Public Sub PrintAnywhere(Src As Object, Dest As Object)
' A Subroutine to Mediate Graphics Presentation upon a Printed Form
    Dest.PaintPicture Src.Picture, Dest.Width / 2, Dest.Height / 2
    If Dest Is Printer Then
      Printer.EndDoc
    End If
    End Sub


    Public Function DSINH(X)
' A Function to Return the Double-Precision Hyperbolic Sine for Argument X
'    Argument:-
'       X     The Double-Precision Argument of DSINH
'
    DSINH = (Exp(X) - Exp(-X)) / 2
    End Function


    Public Function ZETA(ZK, IZ)
' A Function to Return the Double-Precision Riemann's Zeta Function for Argument ZK.
' The Definitional Order(1/n) Algorithm is Employed.
'    Arguments:-
'       ZK    The Double-Precision Argument of ZETA
'       IZ    The Number of Iteration Cycles
'
    ZETA = 0#
    For K = 1 To IZ
      ZETA = ZETA + 1 / K ^ ZK
    Next K
    End Function


    Public Function LOG10(X)
' A Function to Return the Double-Precision Base Ten Logarithm of Arument X
'    Arguments:-
'       X     The Argument whose Base Ten Logarithm is to be Determined
'
    LOG10 = Log(X) / L10
    End Function


    Public Sub OPSIN(NOP, KO(), SOC(), SOP())
' A Subroutine to Input Operations' Descriptors Data
'    Arguments:-
'       NOP   The Number of Operations or Functions
'       KO()  The Array of Operations Serial Numbers
'       SOC() The Array of Operation Codes
```

```
'       SOP()  The Array of Operation Descriptions
' ( Arrays ILogicUnit, STransmode, SPath, SFile and SExt
'   are Common Shared by static Public Declaration )
'
' Open the File
      Call FileOpener(1, IU)
' Input the Data
      Input #IU, NOP
      For I = 1 To NOP
        Input #IU, KO(I), SOC(I), SOP(I)
      Next I
' Terminate
      Close IU
      End Sub

      Public Sub OPSOUT(LNC, NOP, KO(), SOC(), SOP(), LTM())
' A Subroutine to Output Operations' Descriptors Data with
' Operation or Function Timings
'     Arguments:-
'       LNC    The Number of Cycles
'       NOP    The Number of Operations or Functions
'       KO()   The Array of Operations Serial Numbers
'       SOC()  The Array of Operation Codes
'       SOP()  The Array of Operation Descriptions
'       LTM()  The Array of Operation or Function Timings
' ( Arrays ILogicUnit, STransmode, SPath, SFile and SExt
'   are Common Shared by static Public Declaration )
'
' Open the File
      Call FileOpener(2, IV)
' Output the Data
      Write #IV, "Number of Cycles"
      Write #IV, LNC
      Write #IV,
      Write #IV, "Operation Number", "Operation Code", "Operation Description", "Execution Time ( seconds
)"
      For I = 1 To NOP
        Write #IV, KO(I), SOC(I), SOP(I), LTM(I)
      Next I
' Terminate
      Close IV
      End Sub

      Public Sub RANDARG(I, J, A, B, C, D)
' A Subroutine to Generate a Selection of Random Numbers based
' upon the Ranges specified by Arrays IRG() and DRG()
'     Arguments:-
'       I    A Positive Short Integer
'       J    A Negative Short Integer
'       A    A Positive Double-Precision Real Number
'       B    A Positive Double-Precision Real Number
'       C    A Positive Double-Precision Real Number
'       D    A Negative Double-Precision Real Number
' ( Arrays IRG() and DRG() are
'   Common Shared by static Public Declaration )
'
' ** Under Standard Configuration the Random Number Ranges
' ** are as Follows:-
' **
' **    Argument    Lower Bound    Upper Bound
' **    I        1        30
' **    J        1        30
' **    A        0        1
' **    B        0.1       10
```

```
'**      C       0.1     100
'**      D       0       -1000
'**
'
' Compute the Random Integers
    I = IRG(1, 1) + Int((IRG(1, 2) - IRG(1, 1)) * Rnd())
    J = IRG(2, 1) + Int((IRG(2, 2) - IRG(2, 1)) * Rnd())
' Compute the Random Reals
    A = DRG(1, 1) + (DRG(1, 2) - DRG(1, 1)) * Rnd()
    B = DRG(2, 1) + (DRG(2, 2) - DRG(2, 1)) * Rnd()
    C = DRG(3, 1) + (DRG(3, 2) - DRG(3, 1)) * Rnd()
    D = DRG(4, 1) + (DRG(4, 2) - DRG(4, 1)) * Rnd()
' Terminate
    End Sub

    Public Sub OPHOLDERxPWR(KN, LNC, KO(), SOC(), SOP(), LTM())
' A Subroutine to Invoke and Time ann Operation or Function Elaboration
'     Arguments:-
'       KN    The Operation or Function Serial Number
'       LNC   The Number of Cycles
'       NOP   The Number of Operations or Functions
'       KO()  The Array of Operations Serial Numbers
'       SOC() The Array of Operation Codes
'       SOP() The Array of Operation Descriptions
'       LTM() The Array of Operation or Function Timings
'
' Do Composite
    LT1 = GetTickCount()
    For LI = 1 To LNC
      Call OPxPWR
    Next LI
    LT2 = GetTickCount()
' Do Carcass
    LT3 = GetTickCount()
    For LI = 1 To LNC
      Call OPxPWRcarc
    Next LI
    LT4 = GetTickCount()
' Calculate Mill Time
    LTA = LT2 - LT1: LTB = LT4 - LT3
    LTM(KN) = LTA - LTB
' Terminate
    End Sub

    Public Sub OPHOLDERxSQR(KN, LNC, KO(), SOC(), SOP(), LTM())
' A Subroutine to Invoke and Time ann Operation or Function Elaboration
'     Arguments:-
'       KN    The Operation or Function Serial Number
'       LNC   The Number of Cycles
'       NOP   The Number of Operations or Functions
'       KO()  The Array of Operations Serial Numbers
'       SOC() The Array of Operation Codes
'       SOP() The Array of Operation Descriptions
'       LTM() The Array of Operation or Function Timings
'
' Do Composite
    LT1 = GetTickCount()
    For LI = 1 To LNC
      Call OPxSQR
    Next LI
    LT2 = GetTickCount()
' Do Carcass
    LT3 = GetTickCount()
    For LI = 1 To LNC
```

```
        Call OPxSQRcarc
        Next LI
        LT4 = GetTickCount()
' Calculate Mill Time
        LTA = LT2 - LT1: LTB = LT4 - LT3
        LTM(KN) = LTA - LTB
' Terminate
        End Sub

        Public Sub OPHOLDERxMUL(KN, LNC, KO(), SOC(), SOP(), LTM())
' A Subroutine to Invoke and Time ann Operation or Function Elaboration
'       Arguments:-
'       KN      The Operation or Function Serial Number
'       LNC     The Number of Cycles
'       NOP     The Number of Operations or Functions
'       KO()    The Array of Operations Serial Numbers
'       SOC()   The Array of Operation Codes
'       SOP()   The Array of Operation Descriptions
'       LTM()   The Array of Operation or Function Timings
'
' Do Composite
        LT1 = GetTickCount()
        For LI = 1 To LNC
          Call OPxMUL
        Next LI
        LT2 = GetTickCount()
' Do Carcass
        LT3 = GetTickCount()
        For LI = 1 To LNC
          Call OPxMULcarc
        Next LI
        LT4 = GetTickCount()
' Calculate Mill Time
        LTA = LT2 - LT1: LTB = LT4 - LT3
        LTM(KN) = LTA - LTB
' Terminate
        End Sub

        Public Sub OPHOLDERxDIV(KN, LNC, KO(), SOC(), SOP(), LTM())
' A Subroutine to Invoke and Time ann Operation or Function Elaboration
'       Arguments:-
'       KN      The Operation or Function Serial Number
'       LNC     The Number of Cycles
'       NOP     The Number of Operations or Functions
'       KO()    The Array of Operations Serial Numbers
'       SOC()   The Array of Operation Codes
'       SOP()   The Array of Operation Descriptions
'       LTM()   The Array of Operation or Function Timings
'
' Do Composite
        LT1 = GetTickCount()
        For LI = 1 To LNC
          Call OPxDIV
        Next LI
        LT2 = GetTickCount()
' Do Carcass
        LT3 = GetTickCount()
        For LI = 1 To LNC
          Call OPxDIVcarc
        Next LI
        LT4 = GetTickCount()
' Calculate Mill Time
        LTA = LT2 - LT1: LTB = LT4 - LT3
        LTM(KN) = LTA - LTB
```

```
' Terminate
     End Sub

     Public Sub OPHOLDERxADD(KN, LNC, KO(), SOC(), SOP(), LTM())
' A Subroutine to Invoke and Time ann Operation or Function Elaboration
'     Arguments:-
'     KN     The Operation or Function Serial Number
'     LNC    The Number of Cycles
'     NOP    The Number of Operations or Functions
'     KO()   The Array of Operations Serial Numbers
'     SOC()  The Array of Operation Codes
'     SOP()  The Array of Operation Descriptions
'     LTM()  The Array of Operation or Function Timings
'
' Do Composite
     LT1 = GetTickCount()
     For LI = 1 To LNC
       Call OPxADD
     Next LI
     LT2 = GetTickCount()
' Do Carcass
     LT3 = GetTickCount()
     For LI = 1 To LNC
       Call OPxADDcarc
     Next LI
     LT4 = GetTickCount()
' Calculate Mill Time
     LTA = LT2 - LT1: LTB = LT4 - LT3
     LTM(KN) = LTA - LTB
' Terminate
     End Sub

     Public Sub OPHOLDERxSUB(KN, LNC, KO(), SOC(), SOP(), LTM())
' A Subroutine to Invoke and Time ann Operation or Function Elaboration
'     Arguments:-
'     KN     The Operation or Function Serial Number
'     LNC    The Number of Cycles
'     NOP    The Number of Operations or Functions
'     KO()   The Array of Operations Serial Numbers
'     SOC()  The Array of Operation Codes
'     SOP()  The Array of Operation Descriptions
'     LTM()  The Array of Operation or Function Timings
'
' Do Composite
     LT1 = GetTickCount()
     For LI = 1 To LNC
       Call OPxSUB
     Next LI
     LT2 = GetTickCount()
' Do Carcass
     LT3 = GetTickCount()
     For LI = 1 To LNC
       Call OPxSUBcarc
     Next LI
     LT4 = GetTickCount()
' Calculate Mill Time
     LTA = LT2 - LT1: LTB = LT4 - LT3
     LTM(KN) = LTA - LTB
' Terminate
     End Sub

     Public Sub OPHOLDERxASS(KN, LNC, KO(), SOC(), SOP(), LTM())
' A Subroutine to Invoke and Time ann Operation or Function Elaboration
'     Arguments:-
```

```
'      KN     The Operation or Function Serial Number
'      LNC    The Number of Cycles
'      NOP    The Number of Operations or Functions
'      KO()   The Array of Operations Serial Numbers
'      SOC()  The Array of Operation Codes
'      SOP()  The Array of Operation Descriptions
'      LTM()  The Array of Operation or Function Timings
'
' Do Composite
      LT1 = GetTickCount()
      For LI = 1 To LNC
       Call OPxASS
      Next LI
      LT2 = GetTickCount()
' Do Carcass
      LT3 = GetTickCount()
      For LI = 1 To LNC
       Call OPxASScarc
      Next LI
      LT4 = GetTickCount()
' Calculate Mill Time
      LTA = LT2 - LT1: LTB = LT4 - LT3
      LTM(KN) = LTA - LTB
' Terminate
      End Sub

      Public Sub OPHOLDERxBRT(KN, LNC, KO(), SOC(), SOP(), LTM())
' A Subroutine to Invoke and Time ann Operation or Function Elaboration
'    Arguments:-
'      KN     The Operation or Function Serial Number
'      LNC    The Number of Cycles
'      NOP    The Number of Operations or Functions
'      KO()   The Array of Operations Serial Numbers
'      SOC()  The Array of Operation Codes
'      SOP()  The Array of Operation Descriptions
'      LTM()  The Array of Operation or Function Timings
'
' Do Composite
      LT1 = GetTickCount()
      For LI = 1 To LNC
       Call OPxBRT
      Next LI
      LT2 = GetTickCount()
' Do Carcass
      LT3 = GetTickCount()
      For LI = 1 To LNC
       Call OPxBRTcarc
      Next LI
      LT4 = GetTickCount()
' Calculate Mill Time
      LTA = LT2 - LT1: LTB = LT4 - LT3
      LTM(KN) = LTA - LTB
' Terminate
      End Sub

      Public Sub OPHOLDERxEXP(KN, LNC, KO(), SOC(), SOP(), LTM())
' A Subroutine to Invoke and Time ann Operation or Function Elaboration
'    Arguments:-
'      KN     The Operation or Function Serial Number
'      LNC    The Number of Cycles
'      NOP    The Number of Operations or Functions
'      KO()   The Array of Operations Serial Numbers
'      SOC()  The Array of Operation Codes
'      SOP()  The Array of Operation Descriptions
```

```vb
'          LTM()  The Array of Operation or Function Timings
'
' Do Composite
      LT1 = GetTickCount()
      For LI = 1 To LNC
        Call OPxEXP
      Next LI
      LT2 = GetTickCount()
' Do Carcass
      LT3 = GetTickCount()
      For LI = 1 To LNC
        Call OPxEXPcarc
      Next LI
      LT4 = GetTickCount()
' Calculate Mill Time
      LTA = LT2 - LT1: LTB = LT4 - LT3
      LTM(KN) = LTA - LTB
' Terminate
      End Sub


      Public Sub OPHOLDERxLGN(KN, LNC, KO(), SOC(), SOP(), LTM())
' A Subroutine to Invoke and Time ann Operation or Function Elaboration
'    Arguments:-
'      KN     The Operation or Function Serial Number
'      LNC    The Number of Cycles
'      NOP    The Number of Operations or Functions
'      KO()   The Array of Operations Serial Numbers
'      SOC()  The Array of Operation Codes
'      SOP()  The Array of Operation Descriptions
'      LTM()  The Array of Operation or Function Timings
'
' Do Composite
      LT1 = GetTickCount()
      For LI = 1 To LNC
        Call OPxLGN
      Next LI
      LT2 = GetTickCount()
' Do Carcass
      LT3 = GetTickCount()
      For LI = 1 To LNC
        Call OPxLGNcarc
      Next LI
      LT4 = GetTickCount()
' Calculate Mill Time
      LTA = LT2 - LT1: LTB = LT4 - LT3
      LTM(KN) = LTA - LTB
' Terminate
      End Sub


      Public Sub OPHOLDERxL10(KN, LNC, KO(), SOC(), SOP(), LTM())
' A Subroutine to Invoke and Time ann Operation or Function Elaboration
'    Arguments:-
'      KN     The Operation or Function Serial Number
'      LNC    The Number of Cycles
'      NOP    The Number of Operations or Functions
'      KO()   The Array of Operations Serial Numbers
'      SOC()  The Array of Operation Codes
'      SOP()  The Array of Operation Descriptions
'      LTM()  The Array of Operation or Function Timings
'
' Do Composite
      LT1 = GetTickCount()
      For LI = 1 To LNC
        Call OPxL10
```

```
        Next LI
        LT2 = GetTickCount()
' Do Carcass
        LT3 = GetTickCount()
        For LI = 1 To LNC
          Call OPxL10carc
        Next LI
        LT4 = GetTickCount()
' Calculate Mill Time
        LTA = LT2 - LT1: LTB = LT4 - LT3
        LTM(KN) = LTA - LTB
' Terminate
        End Sub

        Public Sub OPHOLDERxZET(KN, LNC, KO(), SOC(), SOP(), LTM())
' A Subroutine to Invoke and Time ann Operation or Function Elaboration
'     Arguments:-
'       KN    The Operation or Function Serial Number
'       LNC   The Number of Cycles
'       NOP   The Number of Operations or Functions
'       KO()  The Array of Operations Serial Numbers
'       SOC() The Array of Operation Codes
'       SOP() The Array of Operation Descriptions
'       LTM() The Array of Operation or Function Timings
'
' Do Composite
        LT1 = GetTickCount()
        For LI = 1 To LNC
          Call OPxZET
        Next LI
        LT2 = GetTickCount()
' Do Carcass
        LT3 = GetTickCount()
        For LI = 1 To LNC
          Call OPxZETcarc
        Next LI
        LT4 = GetTickCount()
' Calculate Mill Time
        LTA = LT2 - LT1: LTB = LT4 - LT3
        LTM(KN) = LTA - LTB
' Terminate
        End Sub

        Public Sub OPHOLDERxSNH(KN, LNC, KO(), SOC(), SOP(), LTM())
' A Subroutine to Invoke and Time ann Operation or Function Elaboration
'     Arguments:-
'       KN    The Operation or Function Serial Number
'       LNC   The Number of Cycles
'       NOP   The Number of Operations or Functions
'       KO()  The Array of Operations Serial Numbers
'       SOC() The Array of Operation Codes
'       SOP() The Array of Operation Descriptions
'       LTM() The Array of Operation or Function Timings
'
' Do Composite
        LT1 = GetTickCount()
        For LI = 1 To LNC
          Call OPxSNH
        Next LI
        LT2 = GetTickCount()
' Do Carcass
        LT3 = GetTickCount()
        For LI = 1 To LNC
          Call OPxSNHcarc
```

```
        Next LI
        LT4 = GetTickCount()
' Calculate Mill Time
        LTA = LT2 - LT1: LTB = LT4 - LT3
        LTM(KN) = LTA - LTB
' Terminate
        End Sub

        Public Sub OPHOLDERxSRT(KN, LNC, KO(), SOC(), SOP(), LTM())
' A Subroutine to Invoke and Time ann Operation or Function Elaboration
'     Arguments:-
'     KN    The Operation or Function Serial Number
'     LNC   The Number of Cycles
'     NOP   The Number of Operations or Functions
'     KO()  The Array of Operations Serial Numbers
'     SOC() The Array of Operation Codes
'     SOP() The Array of Operation Descriptions
'     LTM() The Array of Operation or Function Timings
'
' Do Composite
        LT1 = GetTickCount()
        For LI = 1 To LNC
         Call OPxSRT
        Next LI
        LT2 = GetTickCount()
' Do Carcass
        LT3 = GetTickCount()
        For LI = 1 To LNC
         Call OPxSRTcarc
        Next LI
        LT4 = GetTickCount()
' Calculate Mill Time
        LTA = LT2 - LT1: LTB = LT4 - LT3
        LTM(KN) = LTA - LTB
' Terminate
        End Sub

        Public Sub OPHOLDERxATN(KN, LNC, KO(), SOC(), SOP(), LTM())
' A Subroutine to Invoke and Time ann Operation or Function Elaboration
'     Arguments:-
'     KN    The Operation or Function Serial Number
'     LNC   The Number of Cycles
'     NOP   The Number of Operations or Functions
'     KO()  The Array of Operations Serial Numbers
'     SOC() The Array of Operation Codes
'     SOP() The Array of Operation Descriptions
'     LTM() The Array of Operation or Function Timings
'
' Do Composite
        LT1 = GetTickCount()
        For LI = 1 To LNC
         Call OPxATN
        Next LI
        LT2 = GetTickCount()
' Do Carcass
        LT3 = GetTickCount()
        For LI = 1 To LNC
         Call OPxATNcarc
        Next LI
        LT4 = GetTickCount()
' Calculate Mill Time
        LTA = LT2 - LT1: LTB = LT4 - LT3
        LTM(KN) = LTA - LTB
' Terminate
```

```
     End Sub

     Public Sub OPHOLDERxCOS(KN, LNC, KO(), SOC(), SOP(), LTM())
' A Subroutine to Invoke and Time ann Operation or Function Elaboration
'    Arguments:-
'      KN    The Operation or Function Serial Number
'      LNC   The Number of Cycles
'      NOP   The Number of Operations or Functions
'      KO()  The Array of Operations Serial Numbers
'      SOC() The Array of Operation Codes
'      SOP() The Array of Operation Descriptions
'      LTM() The Array of Operation or Function Timings
'
' Do Composite
     LT1 = GetTickCount()
     For LI = 1 To LNC
       Call OPxCOS
     Next LI
     LT2 = GetTickCount()
' Do Carcass
     LT3 = GetTickCount()
     For LI = 1 To LNC
       Call OPxCOScarc
     Next LI
     LT4 = GetTickCount()
' Calculate Mill Time
     LTA = LT2 - LT1: LTB = LT4 - LT3
     LTM(KN) = LTA - LTB
' Terminate
     End Sub

     Public Sub OPHOLDERxSIN(KN, LNC, KO(), SOC(), SOP(), LTM())
' A Subroutine to Invoke and Time ann Operation or Function Elaboration
'    Arguments:-
'      KN    The Operation or Function Serial Number
'      LNC   The Number of Cycles
'      NOP   The Number of Operations or Functions
'      KO()  The Array of Operations Serial Numbers
'      SOC() The Array of Operation Codes
'      SOP() The Array of Operation Descriptions
'      LTM() The Array of Operation or Function Timings
'
' Do Composite
     LT1 = GetTickCount()
     For LI = 1 To LNC
       Call OPxSIN
     Next LI
     LT2 = GetTickCount()
' Do Carcass
     LT3 = GetTickCount()
     For LI = 1 To LNC
       Call OPxSINcarc
     Next LI
     LT4 = GetTickCount()
' Calculate Mill Time
     LTA = LT2 - LT1: LTB = LT4 - LT3
     LTM(KN) = LTA - LTB
' Terminate
     End Sub

     Public Sub OPHOLDERxLAC(KN, LNC, KO(), SOC(), SOP(), LTM())
' A Subroutine to Invoke and Time ann Operation or Function Elaboration
'    Arguments:-
'      KN    The Operation or Function Serial Number
```

```
'        LNC    The Number of Cycles
'        NOP    The Number of Operations or Functions
'        KO()   The Array of Operations Serial Numbers
'        SOC()  The Array of Operation Codes
'        SOP()  The Array of Operation Descriptions
'        LTM()  The Array of Operation or Function Timings
'
' Do Composite
        LT1 = GetTickCount()
        For LI = 1 To LNC
          Call OPxLAC
        Next LI
        LT2 = GetTickCount()
' Do Carcass
        LT3 = GetTickCount()
        For LI = 1 To LNC
          Call OPxLACcarc
        Next LI
        LT4 = GetTickCount()
' Calculate Mill Time
        LTA = LT2 - LT1: LTB = LT4 - LT3
        LTM(KN) = LTA - LTB
' Terminate
        End Sub

        Public Sub OPHOLDERxAR1(KN, LNC, KO(), SOC(), SOP(), LTM())
' A Subroutine to Invoke and Time ann Operation or Function Elaboration
'      Arguments:-
'        KN     The Operation or Function Serial Number
'        LNC    The Number of Cycles
'        NOP    The Number of Operations or Functions
'        KO()   The Array of Operations Serial Numbers
'        SOC()  The Array of Operation Codes
'        SOP()  The Array of Operation Descriptions
'        LTM()  The Array of Operation or Function Timings
'
' Do Composite
        LT1 = GetTickCount()
        For LI = 1 To LNC
          Call OPxAR1
        Next LI
        LT2 = GetTickCount()
' Do Carcass
        LT3 = GetTickCount()
        For LI = 1 To LNC
          Call OPxAR1carc
        Next LI
        LT4 = GetTickCount()
' Calculate Mill Time
        LTA = LT2 - LT1: LTB = LT4 - LT3
        LTM(KN) = LTA - LTB
' Terminate
        End Sub

        Public Sub OPHOLDERxAR2(KN, LNC, KO(), SOC(), SOP(), LTM())
' A Subroutine to Invoke and Time ann Operation or Function Elaboration
'      Arguments:-
'        KN     The Operation or Function Serial Number
'        LNC    The Number of Cycles
'        NOP    The Number of Operations or Functions
'        KO()   The Array of Operations Serial Numbers
'        SOC()  The Array of Operation Codes
'        SOP()  The Array of Operation Descriptions
'        LTM()  The Array of Operation or Function Timings
```

```
'
' Do Composite
      LT1 = GetTickCount()
      For LI = 1 To LNC
        Call OPxAR2
      Next LI
      LT2 = GetTickCount()
' Do Carcass
      LT3 = GetTickCount()
      For LI = 1 To LNC
        Call OPxAR2carc
      Next LI
      LT4 = GetTickCount()
' Calculate Mill Time
      LTA = LT2 - LT1: LTB = LT4 - LT3
      LTM(KN) = LTA - LTB
' Terminate
      End Sub

      Public Sub OPxPWR()
' A Subroutine to Compute a Real Power of a Real
'    Arguments:-
' ( Arrays IRG() and DRG() are
'   Common Shared by static Public Declaration )
'
      Call RANDARG(I, J, A, B, C, D)
      X = B ^ C
      End Sub

      Public Sub OPxSQR()
' A Subroutine to Compute a Real Square of a Real Number
'    Arguments:-
' ( Arrays IRG() and DRG() are
'   Common Shared by static Public Declaration )
'
      Call RANDARG(I, J, A, B, C, D)
      X = B ^ 2
      End Sub

      Public Sub OPxMUL()
' A Subroutine to Compute a Product of Two Real Numbers
' ( Arrays IRG() and DRG() are
'   Common Shared by static Public Declaration )
'
'     Call RANDARG(I, J, A, B, C, D)
      B = 5: C = 50
      X = B * C
      End Sub

      Public Sub OPxDIV()
' A Subroutine to Compute a Dividend of Two Real Numbers
' ( Arrays IRG() and DRG() are
'   Common Shared by static Public Declaration )
'
'     Call RANDARG(I, J, A, B, C, D)
      B = 5: C = 50
      X = B / C
      End Sub

      Public Sub OPxADD()
' A Subroutine to Compute a Sum of Two Real Numbers
' ( Arrays IRG() and DRG() are
'   Common Shared by static Public Declaration )
'
```

```
'     Call RANDARG(I, J, A, B, C, D)
      B = 5: C = 50
      X = B + C
      End Sub

      Public Sub OPxSUB()
' A Subroutine to Compute a Difference of Two Real Numbers
' ( Arrays IRG() and DRG() are
'   Common Shared by static Public Declaration )
'
'     Call RANDARG(I, J, A, B, C, D)
      B = 5: C = 50
      X = B - C
      End Sub

      Public Sub OPxASS()
' A Subroutine to Assign a Real Number
' ( Arrays IRG() and DRG() are
'   Common Shared by static Public Declaration )
'
'     Call RANDARG(I, J, A, B, C, D)
      B = 5: C = 50
      X = B
      End Sub

      Public Sub OPxBRT()
' A Subroutine to Bracket a Real Number
' ( Arrays IRG() and DRG() are
'   Common Shared by static Public Declaration )
'
'     Call RANDARG(I, J, A, B, C, D)
      B = 5: C = 50
      X = (B)
      End Sub

      Public Sub OPxEXP()
' A Subroutine to Compute a Napierian Exponent of a Real Number
' ( Arrays IRG() and DRG() are
'   Common Shared by static Public Declaration )
'
      Call RANDARG(I, J, A, B, C, D)
      X = Exp(B)
      End Sub

      Public Sub OPxLGN()
' A Subroutine to Compute a Napierian Logarithm of a Real Number
' ( Arrays IRG() and DRG() are
'   Common Shared by static Public Declaration )
'
      Call RANDARG(I, J, A, B, C, D)
      X = Log(C)
      End Sub

      Public Sub OPxL10()
' A Subroutine to Compute a Radix Ten Logarithm of a Real Number
' ( Arrays IRG() and DRG() are
'   Common Shared by static Public Declaration )
'
      Call RANDARG(I, J, A, B, C, D)
      X = LOG10(C)
      End Sub

      Public Sub OPxZET()
' A Subroutine to Compute a Zeta Function of a Real Number
```

```
' ( Arrays IRG() and DRG() are
'  Common Shared by static Public Declaration )
'
    Call RANDARG(I, J, A, B, C, D)
    X = ZETA(B, 2000)
    End Sub

    Public Sub OPxSNH()
' A Subroutine to Compute a Hyperbolic Sine of a Real Number
' ( Arrays IRG() and DRG() are
'  Common Shared by static Public Declaration )
'
    Call RANDARG(I, J, A, B, C, D)
    X = DSINH(A)
    End Sub

    Public Sub OPxSRT()
' A Subroutine to Compute a Square Root of a Positive Real Number
' ( Arrays IRG() and DRG() are
'  Common Shared by static Public Declaration )
'
    Call RANDARG(I, J, A, B, C, D)
    X = Sqr(B)
    End Sub

    Public Sub OPxATN()
' A Subroutine to Compute an Inverse Tangent of a Real Number
' ( Arrays IRG() and DRG() are
'  Common Shared by static Public Declaration )
'
    Call RANDARG(I, J, A, B, C, D)
    X = Atn(B)
    End Sub

    Public Sub OPxCOS()
' A Subroutine to Compute a Cosine of a Real Number
' ( Arrays IRG() and DRG() are
'  Common Shared by static Public Declaration )
'
    Call RANDARG(I, J, A, B, C, D)
    X = Cos(A)
    End Sub

    Public Sub OPxSIN()
' A Subroutine to Compute a Sine of a Real Number
' ( Arrays IRG() and DRG() are
'  Common Shared by static Public Declaration )
'
    Call RANDARG(I, J, A, B, C, D)
    X = Sin(A)
    End Sub

    Public Sub OPxLAC()
' A Subroutine to Execute a FOR-NEXT Lopp Access
' ( Arrays IRG() and DRG() are
'  Common Shared by static Public Declaration )
'
'   Call RANDARG(I,J,A,B,C,D)
    For I = 1 To 1
     II = I
    Next I
    End Sub

    Public Sub OPxAR1()
```

```
' A Subroutine to Execute a One-Dimensional Real Array Access
' ( Arrays IR1(), IR2(), IRG() and DRG() are
'   Common Shared by static Public Declaration )
'
      Call RANDARG(I, J, A, B, C, D)
      AR1(I) = B
      X = AR1(I)
      End Sub

      Public Sub OPxAR2()
' A Subroutine to Execute a Two-Dimensional Real Array Access
' ( Arrays AR1(), AR2(), IRG() and DRG() are
'   Common Shared by static Public Declaration )
'
      Call RANDARG(I, J, A, B, C, D)
      AR2(I, J) = B
      X = AR2(I, J)
      End Sub

      Public Sub OPxPWRcarc()
' A Subroutine to Compute a Real Power of a Real
' ( Arrays IRG() and DRG() are
'   Common Shared by static Public Declaration )
'
      Call RANDARG(I, J, A, B, C, D)
      X = B
      End Sub

      Public Sub OPxSQRcarc()
' A Subroutine to Compute a Real Square of a Real Number
' ( Arrays IRG() and DRG() are
'   Common Shared by static Public Declaration )
'
      Call RANDARG(I, J, A, B, C, D)
      X = B
      End Sub

      Public Sub OPxMULcarc()
' A Subroutine to Compute a Product of Two Real Numbers
' ( Arrays IRG() and DRG() are
'   Common Shared by static Public Declaration )
'
'     Call RANDARG(I, J, A, B, C, D)
      B = 5: C = 50
      X = B
      End Sub

      Public Sub OPxDIVcarc()
' A Subroutine to Compute a Dividend of Two Real Numbers
' ( Arrays IRG() and DRG() are
'   Common Shared by static Public Declaration )
'
'     Call RANDARG(I, J, A, B, C, D)
      B = 5: C = 50
      X = B
      End Sub

      Public Sub OPxADDcarc()
' A Subroutine to Compute a Sum of Two Real Numbers
' ( Arrays IRG() and DRG() are
'   Common Shared by static Public Declaration )
'
'     Call RANDARG(I, J, A, B, C, D)
      B = 5: C = 50
```

```
        X = B
        End Sub

    Public Sub OPxSUBcarc()
' A Subroutine to Compute a Difference of Two Real Numbers
' ( Arrays IRG() and DRG() are
'   Common Shared by static Public Declaration )
'
'       Call RANDARG(I, J, A, B, C, D)
        B = 5: C = 50
        X = B
        End Sub

    Public Sub OPxASScarc()
' A Subroutine to Assign a Real Number
' ( Arrays IRG() and DRG() are
'   Common Shared by static Public Declaration )
'
'       Call RANDARG(I, J, A, B, C, D)
        B = 5: C = 50
        End Sub

    Public Sub OPxBRTcarc()
' A Subroutine to Bracket a Real Number
' ( Arrays IRG() and DRG() are
'   Common Shared by static Public Declaration )
'
'       Call RANDARG(I, J, A, B, C, D)
        B = 5: C = 50
        X = B
        End Sub

    Public Sub OPxEXPcarc()
' A Subroutine to Compute a Napierian Exponent of a Real Number
' ( Arrays IRG() and DRG() are
'   Common Shared by static Public Declaration )
'
        Call RANDARG(I, J, A, B, C, D)
        X = B
        End Sub

    Public Sub OPxLGNcarc()
' A Subroutine to Compute a Napierian Logarithm of a Real Number
' ( Arrays IRG() and DRG() are
'   Common Shared by static Public Declaration )
'
        Call RANDARG(I, J, A, B, C, D)
        X = C
        End Sub

    Public Sub OPxL10carc()
' A Subroutine to Compute a Radix Ten Logarithm of a Real Number
' ( Arrays IRG() and DRG() are
'   Common Shared by static Public Declaration )
'
        Call RANDARG(I, J, A, B, C, D)
        X = C
        End Sub

    Public Sub OPxZETcarc()
' A Subroutine to Compute a Zeta Function of a Real Number
' ( Arrays IRG() and DRG() are
'   Common Shared by static Public Declaration )
'
```

```
        Call RANDARG(I, J, A, B, C, D)
        X = B
        End Sub

        Public Sub OPxSNHcarc()
' A Subroutine to Compute a Hyperbolic Sine of a Real Number
' ( Arrays IRG() and DRG() are
'   Common Shared by static Public Declaration )
'
        Call RANDARG(I, J, A, B, C, D)
        X = A
        End Sub

        Public Sub OPxSRTcarc()
' A Subroutine to Compute a Square Root of a Positive Real Number
' ( Arrays IRG() and DRG() are
'   Common Shared by static Public Declaration )
'
        Call RANDARG(I, J, A, B, C, D)
        X = B
        End Sub

        Public Sub OPxATNcarc()
' A Subroutine to Compute an Inverse Tangent of a Real Number
' ( Arrays IRG() and DRG() are
'   Common Shared by static Public Declaration )
'
        Call RANDARG(I, J, A, B, C, D)
        X = B
        End Sub

        Public Sub OPxCOScarc()
' A Subroutine to Compute a Cosine of a Real Number
' ( Arrays IRG() and DRG() are
'   Common Shared by static Public Declaration )
'
        Call RANDARG(I, J, A, B, C, D)
        X = A
        End Sub

        Public Sub OPxSINcarc()
' A Subroutine to Compute a Sine of a Real Number
' ( Arrays IRG() and DRG() are
'   Common Shared by static Public Declaration )
'
        Call RANDARG(I, J, A, B, C, D)
        X = A
        End Sub

        Public Sub OPxLACcarc()
' A Subroutine to Execute a FOR-NEXT Loop Access
' ( Arrays IRG() and DRG() are
'   Common Shared by static Public Declaration )
'
'       Call RANDARG(I,J,A,B,C,D)
'       For I = 1 To 1
          II = I
'       Next I
        End Sub

        Public Sub OPxAR1carc()
' A Subroutine to Execute a One-Dimensional Real Array Access
'   Common Shared by static Public Declaration )
'
```

```
    Call RANDARG(I, J, A, B, C, D)
    AR1(I) = B
    X = B
    End Sub


    Public Sub OPxAR2carc()
' A Subroutine to Execute a Two-Dimensional Real Array Access
' ( Arrays AR1(), AR2(), IRG() and DRG() are
'   Common Shared by static Public Declaration )
'

    Call RANDARG(I, J, A, B, C, D)
    AR2(I, J) = B
    X = B
    End Sub
```

APPENDIX F

The Timing Tariff Results Worksheet

Operation Serial	Operation Code	Operation Description	Log$_{10}$(Mean in Nanoseconds)	Mean (Nanoseconds)	Mean	Population Standard Deviation	Coefficient of Variation
1 PWR	B^C		6.1580	472.4951	4.724951E-07	6.13931E-08	12.99
2 SQR	B^2		6.1723	479.3023	4.793023E-07	5.31871E-08	11.10
3 MUL	B*C		2.9817	19.7222	1.972222E-08	1.90216E-08	96.45
4 DIV	B/C		4.4217	83.2346	8.323464E-08	1.12404E-07	135.05
5 ADD	B+C		3.5926	36.3278	3.632778E-08	4.4224E-08	121.74
6 SUB	B-C		3.6509	38.5087	3.850868E-08	4.52378E-08	117.47
7 ASS	B:=C		3.7724	43.4826	4.348264E-08	4.31338E-08	99.20
8 BRT	(B)		2.9787	19.6624	1.966239E-08	2.8428E-08	144.58
9 EXP	EXP(B)		6.0523	425.1078	4.251078E-07	1.31993E-07	31.05
10 LGN	LOGN(C)		4.9817	145.7188	1.457188E-07	5.70032E-08	39.12
11 L10	LOG10(C)		6.0817	437.7876	4.377876E-07	1.27637E-07	29.15
12 ZET	ZETA(B)		13.9025	1090891.0980	1.090891E-03	7.87264E-05	7.22
13 SNH	SINH(A)		6.9493	1042.4134	1.042413E-06	2.06026E-07	19.76
14 SRT	SQRT(B)		4.1034	60.5469	6.054688E-08	3.49183E-08	57.67
15 ATN	ATAN(B)		5.1669	175.3681	1.753681E-07	6.47233E-08	36.91
16 COS	COS(A)		4.7877	120.0226	1.200226E-07	4.52516E-08	37.70
17 SIN	SIN(A)		5.0014	148.6250	1.486250E-07	4.84302E-08	32.59
18 LAC	Loop Access		5.0230	151.8611	1.518611E-07	9.13211E-08	60.13
19 AR1	1D Real Array Access		4.9726	144.3981	1.443981E-07	2.51902E-07	174.45
20 AR2	2D Real Array Access		4.4909	89.2031	8.920313E-08	5.7651E-08	64.63

Operation Serial	Operation Code	Operation Description	TG10000	TG25000	TG50000	TG100000	TG100000A
FileName			TG10000	TG25000	TG50000	TG100000	TG100000A
Iterations			10000	25000	50000	100000	100000
Negatives			0	0	0	0	0
Totals Seconds			11.48	28.73	57.45	114.96	114.96
Operations Seconds			10.68	26.68	53.32	106.72	106.65
Percent Time for Operations			93.01393728	92.86112078	92.818027	92.83141962	92.77487822
1	PWR	B^C	3.000000E-07	5.200000E-07	4.800000E-07	4.600000E-07	4.100000E-07
2	SQR	B^2	5.000000E-07	4.000000E-07	5.000000E-07	5.000000E-07	4.000000E-07
3	MUL	B*C		0.000000E+00	0.000000E+00	0.000000E+00	0.000000E+00
4	DIV	B/C	5.000000E-07	2.000000E-07	0.000000E+00	1.000000E-08	1.000000E-07
5	ADD	B+C	0.000000E+00	2.000000E-07	0.000000E+00	0.000000E+00	5.000000E-08
6	SUB	B-C		2.000000E-07	0.000000E+00	0.000000E+00	0.000000E+00
7	ASS	B:=C			1.000000E-07	5.000000E-08	5.000000E-08
8	BRT	(B)				5.000000E-08	0.000000E+00
9	EXP	EXP(B)	5.000000E-07	4.000000E-07	9.000000E-07	3.500000E-07	3.500000E-07
10	LGN	LOGN(C)	0.000000E+00		1.000000E-07	5.000000E-08	2.500000E-07
11	L10	LOG10(C)	5.000000E-07	4.000000E-07	4.000000E-07	4.000000E-07	4.000000E-07
12	ZET	ZETA(B)	1.063500E-03	1.063000E-03	1.062300E-03	1.063520E-03	1.062800E-03
13	SNH	SINH(A)	1.000000E-06	1.000000E-06	1.000000E-06	1.000000E-06	1.000000E-06
14	SRT	SQRT(B)	0.000000E+00	0.000000E+00	1.000000E-07	1.000000E-07	4.000000E-08
15	ATN	ATAN(B)		4.000000E-08	1.000000E-07	1.500000E-07	1.400000E-07
16	COS	COS(A)	0.000000E+00	2.000000E-07	1.000000E-07	1.500000E-07	1.000000E-07
17	SIN	SIN(A)	0.000000E+00	2.000000E-07	2.000000E-07	1.000000E-07	1.500000E-07
18	LAC	Loop Access	5.000000E-07	2.000000E-07	1.000000E-07	1.500000E-07	1.000000E-07
19	AR1	1D Real Array Access	5.000000E-07		0.000000E+00	1.500000E-07	1.500000E-07
20	AR2	2D Real Array Access	0.000000E+00	0.000000E+00	1.000000E-07	1.500000E-07	5.000000E-08

FileName	TG150000	TG150000A	TG160000	TG180000	TG200000	TG500000	TG1000000
Iterations	150000	150000	160000	180000	200000	500000	1000000
Negatives	0	0	0	0	0	0	0
Totals Seconds	172.41	172.36	183.95	206.91	229.92	574.69	1149.42
Operations Seconds	159.94	159.94	170.73	191.95	213.30	533.41	1066.73
Percent Time for Operations	92.76666087	92.79125087	92.81217722	92.76787009	92.77009395	92.81612696	92.80550191

Operation Serial	Operation Code	Operation Description	TG150000	TG150000A	TG160000	TG180000	TG200000	TG500000	TG1000000
1	PWR	B^C	5.666667E-07	4.866667E-07	4.562500E-07	3.833333E-07	5.050000E-07	4.800000E-07	4.750000E-07
2	SQR	B^2	4.666667E-07	5.666667E-07	4.687500E-07	4.722222E-07	4.000000E-07	4.700000E-07	4.750000E-07
3	MUL	B*C	6.666667E-08	0.000000E+00	0.000000E+00	0.0000000E+00	3.500000E-08	2.000000E-08	3.000000E-08
4	DIV	B/C	3.333333E-08	3.333333E-08	3.125000E-08	5.555556E-08	7.500000E-08	6.000000E-08	5.000000E-08
5	ADD	B+C	3.333333E-08	0.000000E+00	3.125000E-08	5.555556E-08	2.500000E-08	4.000000E-08	2.500000E-08
6	SUB	B-C	0.000000E+00	3.333333E-08	3.125000E-08	5.555556E-08	5.000000E-08	3.000000E-08	3.000000E-08
7	ASS	B=C	3.333333E-08	0.000000E+00	3.125000E-08	5.555556E-08	5.000000E-08	3.000000E-08	3.500000E-08
8	BRT	(B)	3.333333E-08	3.333333E-08	0.000000E+00	2.77778E-08			5.000000E-09
9	EXP	EXP(B)	3.333333E-07	4.000000E-07	4.062500E-07	3.333333E-07	3.500000E-07	3.900000E-07	3.850000E-07
10	LGN	LOGN(C)	1.666667E-07	1.666667E-07	1.875000E-07	1.666667E-07	1.250000E-07	1.200000E-07	1.550000E-07
11	L10	LOG10(C)	3.333333E-07	4.000000E-07	4.062500E-07	3.055556E-07	3.000000E-07	3.800000E-07	3.900000E-07
12	ZET	ZETA(B)	1.062567E-03	1.062500E-03	1.063156E-03	1.062500E-03	1.063000E-03	1.063050E-03	1.063015E-03
13	SNH	SINH(A)	9.333333E-07	9.333333E-07	1.000000E-06	9.44444E-07	9.000000E-07	9.700000E-07	9.800000E-07
14	SRT	SQRT(B)	3.333333E-08	1.000000E-07	3.125000E-08	6.666667E-08	5.000000E-08	5.000000E-08	3.000000E-08
15	ATN	ATAN(B)	1.666667E-07	1.666667E-07	2.812500E-07	2.222222E-07	1.750000E-07	1.500000E-07	1.450000E-07
16	COS	COS(A)	1.000000E-07	1.000000E-07	1.250000E-07	1.111111E-07	1.000000E-07	9.000000E-08	1.150000E-07
17	SIN	SIN(A)	1.066667E-07	1.666667E-07	1.562500E-07	1.666667E-07	1.200000E-07	1.600000E-07	1.450000E-07
18	LAC	Loop Access	1.200000E-07	1.000000E-07	9.375000E-08	1.388889E-07	1.100000E-07	1.400000E-07	1.200000E-07
19	AR1	1D Real Array Access	6.666667E-08	6.666667E-09	6.250000E-08	5.555556E-08	6.500000E-08	8.000000E-08	3.000000E-08
20	AR2	2D Real Array Access	1.000000E-07	4.000000E-08	9.375000E-08	2.500000E-07	5.000000E-08	1.000000E-07	9.000000E-08

Operation Serial	Operation Code	Operation Description	TH1M	TH2M	TH25MA	TH3M	TH4M
		FileName	TH1M	TH2M	TH25MA	TH3M	TH4M
		Iterations	1000000	2000000	2500000	3000000	4000000
		Negatives	0	0	0	0	0
		Totals Seconds	1147.67	2296.05	2875.57	4297.64	5615.09
		Operations Seconds	1063.99	2128.64	2655.46	3971.58	5194.95
		Percent Time for Operations	92.7087926	92.70891313	92.34544803	92.41311511	92.51769785
1 PWR		B^C	4.560000E-07	5.115000E-07	4.880000E-07	4.990000E-07	5.550000E-07
2 SQR		B^2	4.900000E-07	4.550000E-07	4.880000E-07	4.783333E-07	6.175000E-07
3 MUL		B*C	2.000000E-08	3.250000E-08	3.000000E-08	2.666667E-08	3.500000E-08
4 DIV		B/C	5.000000E-08	4.500000E-08	5.360000E-08	5.166667E-08	6.625000E-08
5 ADD		B+C	3.500000E-08	4.000000E-08	2.160000E-08	3.333333E-08	2.750000E-08
6 SUB		B-C	3.500000E-08	3.500000E-08	3.600000E-08	3.500000E-08	4.500000E-08
7 ASS		B:=C	3.000000E-08	3.500000E-08	3.600000E-08	3.333333E-08	2.625000E-08
8 BRT		(B)	0.000000E+00	0.000000E+00	2.000000E-09	1.666667E-09	0.000000E+00
9 EXP		EXP(B)	4.050000E-07	4.525000E-07	3.460000E-07	3.716667E-07	5.537500E-07
10 LGN		LOGN(C)	1.350000E-07	1.625000E-07	1.840000E-07	1.800000E-07	1.825000E-07
11 L10		LOG10(C)	6.200000E-07	8.500000E-07	4.160000E-07	4.150000E-07	5.262500E-07
12 ZET		ZETA(B)	1.059815E-03	1.059898E-03	1.058232E-03	1.318430E-03	1.293866E-03
13 SNH		SINH(A)	1.000000E-06	1.015000E-06	9.920000E-07	1.821667E-06	1.231250E-06
14 SRT		SQRT(B)	1.000000E-07	8.500000E-08	8.000000E-08		1.025000E-07
15 ATN		ATAN(B)	1.850000E-07	1.650000E-07	1.820000E-07	3.333333E-07	2.037500E-07
16 COS		COS(A)	1.650000E-07	1.425000E-07	1.280000E-07		1.937500E-07
17 SIN		SIN(A)	1.800000E-07	1.500000E-07	1.780000E-07		1.987500E-07
18 LAC		Loop Access	1.150000E-07	1.275000E-07	1.340000E-07	1.700000E-07	1.625000E-07
19 AR1		1D Real Array Access	5.000000E-08	4.250000E-08	5.800000E-08	9.803333E-08	1.875000E-08
20 AR2		2D Real Array Access	1.050000E-07	7.500000E-08	9.800000E-08		1.255000E-07

APPENDIX G

Zipfian Plots and Tabulations

		DP	DR	ES	EP	LA	HT	ST	SE	WO	WT	BS	WS	All Classical Group	All Lanczos Group	All Fitted Polynomial Group	All Stirling Group	All Methods
1 PWR	B^C	26	0	18	0	1	9	1	8	1	5	1	2	44	1	9	18	72
2 SQR	B^2	0	0	0	0	0	0	0	0	0	0	0	0	0	0	0	0	0
3 MUL	B*C	26	52	20	23	9	9	2	3	7	12	2	4	121	9	9	30	169
4 DIV	B/C	1	1	10	21	13	0	0	7	2	6	1	4	33	13	0	20	66
5 ADD	B+C	26	26	9	10	10	10	0	7	1	9	2	1	71	10	10	20	111
6 SUB	B-C	0	0	1	0	7	9	2	9	3	3	2	0	1	7	9	19	36
7 ASS	B:=C	28	55	20	12	8	11	1	9	1	6	1	1	115	8	11	19	153
8 BRT	(B)	0	0	0	11	7	9	0	15	5	9	3	4	31	7	9	36	83
9 EXP	EXP(B)	0	0	1	11	1	0	1	1	1	1	1	0	12	1	0	5	18
10 LGN	LOGN(C)	0	0	0	0	0	0	0	0	0	0	0	0	0	0	0	0	0
11 L10	LOG10(C)	0	0	0	0	0	0	0	0	0	0	0	0	0	0	0	0	0
12 ZET	ZETA(B)	0	0	9	0	0	0	0	0	0	0	0	0	9	0	0	0	9
13 SNH	SINH(A)	0	0	0	0	0	0	0	0	0	0	0	1	0	0	0	1	1
14 SRT	SQRT(B)	0	0	0	0	0	0	0	0	0	0	0	2	0	0	0	2	2
15 ATN	ATAN(B)	0	0	0	0	0	0	0	0	0	0	0	0	0	0	0	0	0
16 COS	COS(A)	0	0	0	0	0	0	0	0	0	0	0	0	0	0	0	0	0
17 SIN	SIN(A)	0	0	0	0	0	0	0	0	0	0	0	0	0	0	0	0	0
18 LAC	Loop Access	26	26	9	10	6	9	0	7	0	4	0	0	71	6	9	11	97
19 AR1	1D Real Array Access	0	0	0	0	0	0	0	0	0	0	0	0	0	0	0	0	0
20 AR2	2D Real Array Access	26	26	0	0	7	9	0	7	0	4	0	0	52	7	9	11	79

All Methods Rank versus Log$_n$(n)

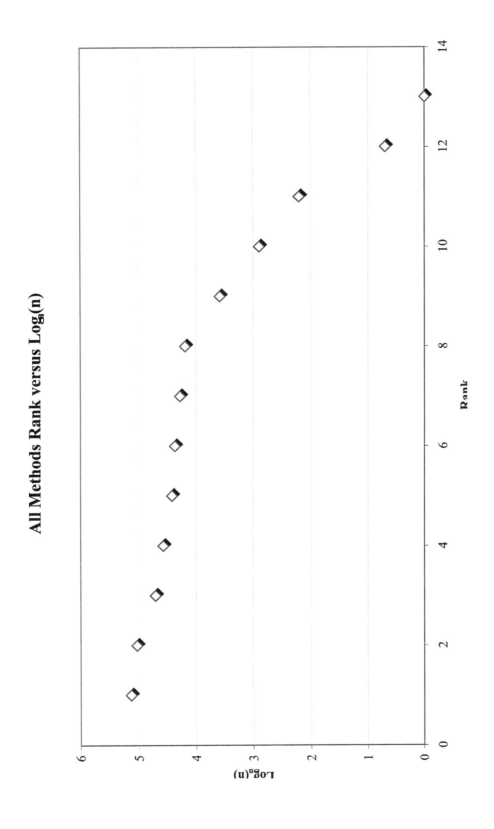

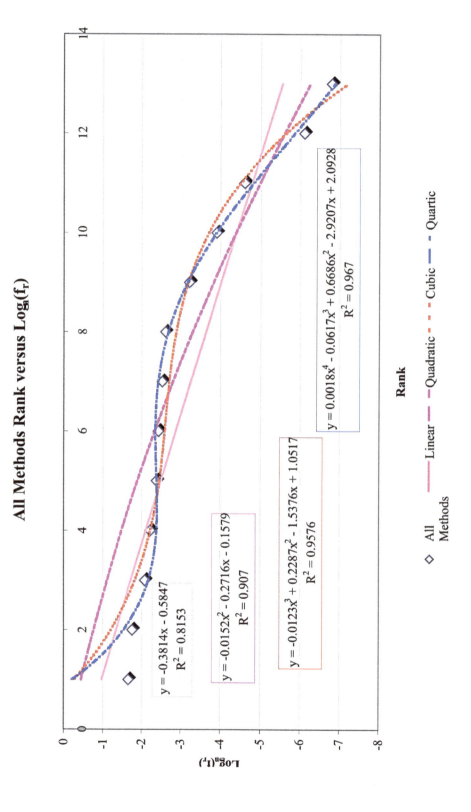

All Methods Rank versus $Log_{sn}(f_r)$

$y = -0.3814x - 0.5847$
$R^2 = 0.8153$

$y = -0.0152x^2 - 0.2716x - 0.1579$
$R^2 = 0.907$

$y = -0.0123x^3 + 0.2287x^2 - 1.5376x + 1.0517$
$R^2 = 0.9576$

$y = 0.0018x^4 - 0.0617x^3 + 0.6686x^2 - 2.9207x + 2.0928$
$R^2 = 0.967$

Rank

◇ All Methods ── Linear ── Quadratic ── Cubic ── Quartic

$Log_{sn}(f_r)$

A 14.14829898
B 3.605344514
C 3614.955309

Data

r	Logn(1/f)	1/fr	fr_actual	fr_modelled
7	2.521274	12.444444	0.080357	0.060260
1	1.668042	5.301775	0.18616	0.200671
8	2.608286	13.575758	0.073661	0.051014
3	2.088410	8.072072	0.123884	0.128325
9	3.214421	24.888889	0.040179	0.043505
2	1.767502	5.856209	0.170759	0.159364
5	2.379100	10.795181	0.092634	0.086215
10	3.907569	49.777778	0.020089	0.037353
11	4.600716	99.555556	0.010045	0.032269
13	6.797940	896.000000	0.001116	0.024489
12	6.104793	448.000000	0.002232	0.028037
4	2.223229	9.237113	0.108259	0.104608
6	2.428493	11.341772	0.088170	0.071759

Sorted Plotting Gatherment

δ^2	r	fr_actual	fr_modelled
0.000404	1	0.188616071	0.200671
0.000145	2	0.170758929	0.159364
0.000513	3	0.123883929	0.128325
0.000020	4	0.108258929	0.104608
0.000011	5	0.092633929	0.086215
0.000130	6	0.088169643	0.071759
0.000041	7	0.080357143	0.060260
0.000298	8	0.073660714	0.051014
0.000494	9	0.040178571	0.043505
0.000546	10	0.020089286	0.037353
0.000666	11	0.010044643	0.032269
0.000013	12	0.002232143	0.028037
0.000269	13	0.001116071	0.024489

General Zipfs Law Model for All Methods

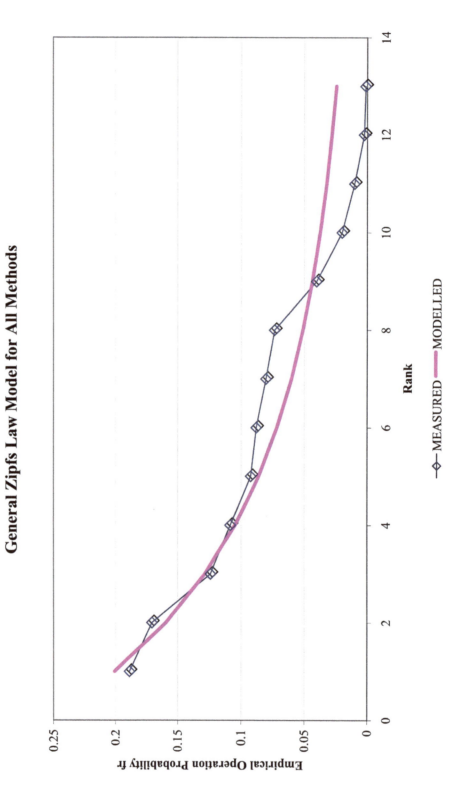

r	Logn(1/f)
7	2.521274
1	1.668042
8	2.608286
3	2.088410
9	3.214421
2	1.767502
5	2.379100
10	3.907569
11	4.600716
13	6.797940
12	6.104793
4	2.223229
6	2.428493
	0.584655
	0.381420
	0.815307

r	Logn(1/f)
1	1.668042
2	1.767502
3	2.088410
4	2.223229
5	2.379100
6	2.428493
7	2.521274
8	2.608286
9	3.214421
10	3.907569
11	4.600716
12	6.104793
13	6.797940
	0.584655
	0.381420
	0.815307

r	Logn(1/f)
1	1.668042
2	1.767502
3	2.088410
4	2.223229
5	2.379100
6	2.428493
7	2.521274
8	2.608286
	1.593041
	0.137222
	0.945624

r	Logn(1/f)
8	2.608286
9	3.214421
10	3.907569
11	4.600716
12	6.104793
13	6.797940
	-4.554807
	0.866072
	0.976777

NTERCEPT:
iRADIENT:
)ETERMINATION COEFFICIENT:

Segmental Zipfian Plot for All Methods

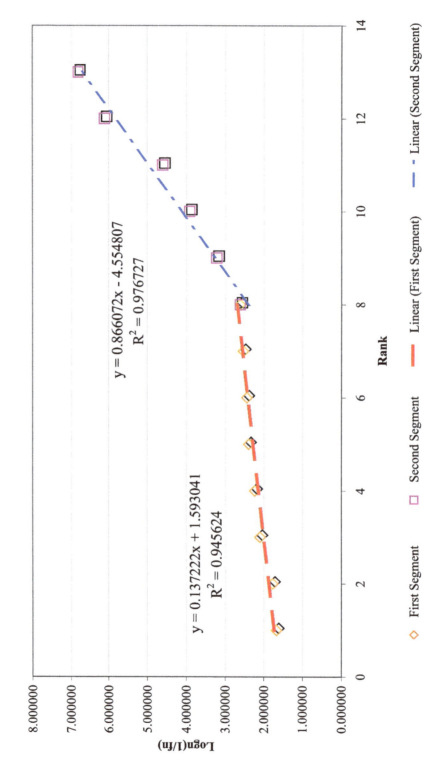

$y = 0.866072x - 4.554807$
$R^2 = 0.976727$

$y = 0.137222x + 1.593041$
$R^2 = 0.945624$

Rank

Logn(1/fn)

◇ First Segment   □ Second Segment   — Linear (First Segment)   – · Linear (Second Segment)

APPENDIX H

Temporal Ideality Profiles

Operation Serial	Operation Code	Operation Description	Mean (Nanoseconds)	Time Reciprocal	Time Fraction	DP	DR	ES	EP	LA	HT	ST	SE	WO	WT	BS	WS	All	Classical Group IV
1	PWR	B^C	472.4951	0.0021164241	0.0078756162	0.0313199		0.0210526		0.3324167	0.1295642		0.25	0.0539107	0.1931923	0.0748441	0.1714286	0.1052632	0.016909
2	SQR	B^2	479.3023	0.0020863660	0.0077637644														
3	MUL	B*C	19.7222	0.0507042254	0.1886800587	0.0313199	0.0166818	0.0189474	0.0781633	0.0369352	0.1295642		0.125	0.1437618	0.031185	0.0857143	0.0526316		0.0061487
4	DIV	B/C	83.2346	0.0120142286	0.0447072279	0.8143177	0.8674553	0.0378947	0.0856075	0.0255705	0.1166078			0.0616122	0.0965961	0.0623701	0.1714286	0.0526316	0.0225453
5	ADD	B+C	36.3278	0.0275271448	0.1024337373	0.0313199	0.0333637	0.0421053	0.1797757	0.0332417	0.1295642		0.125	0.0616122	0.1931923	0.1247401	0.0857143	0.2105263	0.0104788
6	SUB	B-C	38.5087	0.0259681710	0.0966324993	0.0290828	0.0157719	0.3789474		0.0474881	0.1295642		0.125	0.0479206	0.0643974	0.1247401	0.0857143	0.0857143	0.7439963
7	ASS	B:=C	43.4826	0.0229976843	0.0855787538			0.0189474	0.1498131	0.0415221	0.1060071		0.25	0.0479206	0.1931923	0.0643974	0.2105263	0.2105263	0.0064695
8	BRT	(B)	19.6624	0.0508585090	0.1892541776			0.0189474	0.1634324	0.0474881	0.1295642		0.25	0.0287524	0.0386385	0.0571429	0.0526316		0.0239999
9	EXP	EXP(B)	425.1078	0.0023523443	0.0087535201			0.3789474	0.1634324	0.3324167			0.25	0.4312853	0.1931923	0.3742204	0.1714286		0.0619997
10	LGN	LOGN(C)	145.7188	0.0068625348	0.0255367964														
11	L10	LOG10(C)	437.7876	0.0022842128	0.0084999900														
12	ZET	ZETA(B)	1090891.0980	0.0000009167	0.0000034111			0.0421053											0.0826663
13	SNH	SINH(A)	1042.4134	0.0009593123	0.0035697834													0.2105263	
14	SRT	SQRT(B)	60.5469	0.0165161290	0.0614596550													0.1052632	
15	ATN	ATAN(B)	175.3681	0.0057022928	0.0212193152														
16	COS	COS(A)	120.0226	0.0083317663	0.0310040859														
17	SIN	SIN(A)	148.6250	0.0067283431	0.0250374435														
18	LAC	Loop Access	151.8611	0.0065849643	0.0245039037	0.0313199	0.0333637	0.0421053	0.1797757	0.0554028	0.1295642		0.0616122		0.0935551				0.0104788
19	AR1	1D Real Array Access	144.3981	0.0069252966	0.0257703447										0.0935551				0.0104788
20	AR2	2D Real Array Access	89.2031	0.0112103696	0.0417159158	0.0313199	0.0333637			0.0474881	0.1295642		0.0616122		0.0935551				0.0143076
		TOTAL	1095024.8862	0.2687312358	1.0000000000	1	1	1	1	1	1		1	1	1	1	1	1	1

Method Ideality Profiles for Time
(Squared Deviations)

Method Ideality Profiles for Time
(Deviations)

Method Departure from Temporal Ideality
in terms of RMS(Time Fraction Deviation)

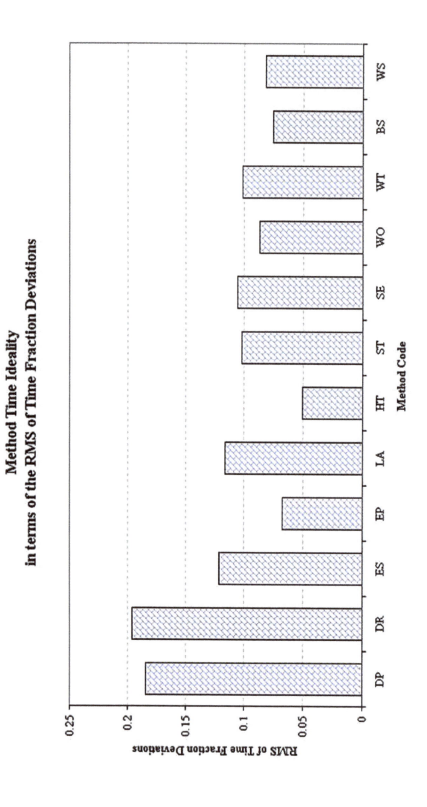

Method Time Ideality
in terms of the RMS of Time Fraction Deviations

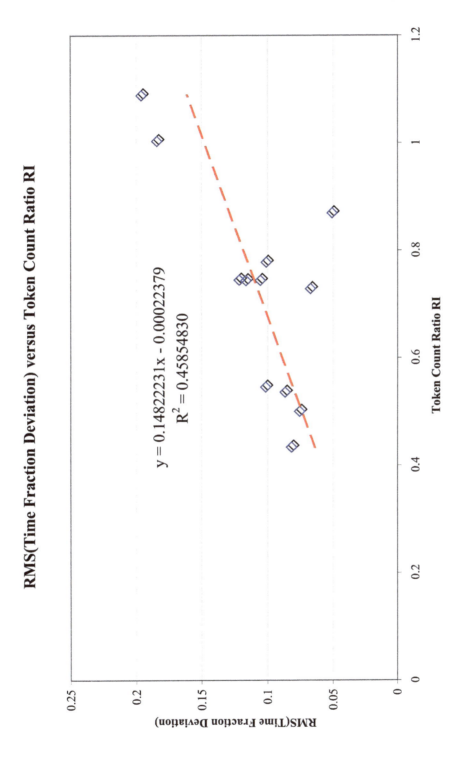

RMS(Time Fraction Deviation) versus Token Count Ratio RI

$y = 0.14822231x - 0.00022379$
$R^2 = 0.45854830$

Token Count Ratio RI

RMS(Time Fraction Deviation)

www.ingramcontent.com/pod-product-compliance
Lightning Source LLC
LaVergne TN
LVHW061953050326
832904LV00010B/299